Schwarz's Lemma from a Differential Geometric Viewpoint

IISc LECTURE NOTES SERIES

ISSN: 2010-2402

IISc Lecture Notes Series

Schwarz's Lemma from a Differential Geometric Viewpoint

Kang-Tae Kim

Pohang University of Science and Technology, Korea

Hanjin Lee

Handong Global University, Korea

IISc Press

World Scientific

NEW JERSEY · LONDON · SINGAPORE · BEIJING · SHANGHAI · HONG KONG · TAIPEI · CHENNAI

Published by

World Scientific Publishing Co. Pte. Ltd.

5 Toh Tuck Link, Singapore 596224

USA office: 27 Warren Street, Suite 401-402, Hackensack, NJ 07601

UK office: 57 Shelton Street, Covent Garden, London WC2H 9HE

British Library Cataloguing-in-Publication Data
A catalogue record for this book is available from the British Library.

SCHWARZ'S LEMMA FROM A DIFFERENTIAL GEOMETRIC VIEWPOINT
IISc Lecture Notes Series — Vol. 2

ISBN-13 978-981-4324-78-6
ISBN-10 981-4324-78-7

Printed in Singapore.

To our mothers

Series Preface

World Scientific Publishing Company - Indian Institute of Science Collaboration

IISc Press and WSPC are co-publishing books authored by world renowned scientists and engineers. This collaboration, started in 2008 during IISc's centenary year under a Memorandum of Understanding between IISc and WSPC, has resulted in the establishment of three Series: IISc Centenary Lectures Series (ICLS), IISc Research Monographs Series (IRMS), and IISc Lecture Notes Series (ILNS).

This pioneering collaboration will contribute significantly in disseminating current Indian scientific advancement worldwide.

The **"IISc Centenary Lectures Series"** will comprise lectures by designated Centenary Lecturers - eminent teachers and researchers from all over the world.

The **"IISc Research Monographs Series"** will comprise state-of-the-art monographs written by experts in specific areas. They will include, but not limited to, the authors' own research work.

The **"IISc Lecture Notes Series"** will consist of books that are reasonably self-contained and can be used either as textbooks or for self-study at the postgraduate level in science and engineering. The books will be based on material that has been class-tested for most part.

Editorial Board for the IISc Lecture Notes Series (ILNS):

Gadadhar Misra, Editor-in-Chief (gm@math.iisc.ernet.in)

Chandrashekar S Jog (jogc@mecheng.iisc.ernet.in)
Joy Kuri (kuri@cedt.iisc.ernet.in)
K L Sebastian (kls@ipc.iisc.ernet.in)
Diptiman Sen (diptiman@cts.iisc.ernet.in)
Sandhya Visweswariah (sandhya@mrdg.iisc.ernet.in)

Preface

Schwarz's Lemma was only a small preparatory lemma at its initial discovery around 1880. (It seems difficult to locate its historical origin precisely; see [Osserman 1999a].) But it has grown so much that it has become a principal tool in various branches of mathematics. We were astonished when the internet search showed more than 150 "hits" (i.e., 150 research papers detected in the search) with the keyword "Schwarz's Lemma". As much as it was a surprise to us, we were inspired to work on a survey, mostly for our own sake in the first place, of the stream of research developments pertaining to Schwarz's Lemma and its developments.

It did not take us too long to realize, after some reading, that there are indeed quite a few but not too many fundamental achievements that provide "core ideas and methods". Some of them, however, might not be so easy to understand in a quick reading. Thus we were convinced that it might be worthwhile to write these notes on its differential geometric developments in their present form.

These notes are of a classical nature and start with the original Schwarz's Lemma—preceded only by some preliminaries on harmonic and subharmonic functions (Chapter 1). The modification by Pick (around 1916; see [Kobayashi 1970]), now known as the Schwarz-Pick Lemma, is introduced in the same chapter and is re-interpreted in terms of the Poincaré metric and distance. This establishes the beginning stage of differential geometric ideas making bridges with Schwarz's Lemma. These matters constitute Chapter 2.

In the 1930's, the statement *"Negative curvature restricts holomorphic mappings"* emerged as an important slogan in research on differential geometry and geometric (complex) analysis. The generalizations of Schwarz's Lemma from the viewpoint of differential geometry involving curvature have obviously played an important role in research up to the present days. A remarkably well-written survey article by Osserman ([Osserman 1999a]) provides a good historical account.

It is widely agreed that the generalization by Ahlfors (in 1938) of Schwarz's Lemma ([Ahlfors 1938]) was the key result that opened the first door to the subsequent developments. Ahlfors investigated the holomorphic mappings from the unit disc into a Riemann surface that admits a Hermitian metric with its curvature bounded from above by a negative constant. He obtained the upper bound estimate of the pull back, by the holomorphic map, of the Hermitian metric tensor by the Poincaré metric tensor up to a constant multiple; the multiplier is the quotient, that is the curvature of the Poincaré metric of the source disc of the map divided by the negative upper bound of the curvature of the Hermitian metric of the target Riemann surface.

As described in Chapter 3, Ahlfors' generalization and proof can be viewed as follows: For a Riemann surface M with a Hermitian metric ds^2_M and a holomorphic mapping $f : D \to M$ from the open unit disc D into M, the pull-back $f^*ds^2_M$ is a non-negative Hermitian symmetric tensor on D. Since D is complex one-dimensional, any Hermitian symmetric $(1,1)$-tensor is a scalar multiple of the other. Therefore $f^*ds^2_M - u\, ds^2_D$, where ds^2_D is the Poincaré metric of D and u is a non-negative real-valued function on D.

In case u attains its maximum, say at z_0, it suffices to show that $u(z_0)$ is bounded from above by the quotient of two curvature bounds. If $u(z_0) = 0$, then there is nothing more to prove. If $u(z_0) > 0$, then one has $\nabla \log u|_{z_0} = 0$, and $\Delta \log u|_{z_0} \leq 0$. With the definition of u and the curvatures of metrics involved, this (after some clever calculations) yields the bound for u at the maximum point z_0, and hence the desired upper bound estimate for u at every point by the ratio of curvature bounds.

But there is no guarantee in general that u should attain its maximum in D. So Ahlfors introduced a technique of "shrinking the disc" which ensured the existence of maximum for the multiplier function u on the shrunken disc. (See Chapter 3 for details.) Then letting the shrunken disc to expand back to D, the intermediate estimates then yield the desired conclusion (at the limit).

In order to go beyond the complex one-dimensional manifolds, it turns out that the major barriers seem residing with the high dimensionality of the domain manifolds. In high dimensions, one has to understand how to compare two Hermitian tensors up to scalar function multipliers. And, even after that step is successfully carried out, one still needs some method to relate the upper bound estimate for the multiplier by the ratio of curvatures. These problems were successfully resolved by S.S. Chern ([Chern 1968]) and Y.C. Lu ([Lu 1968]). That work involves much of the concepts and methods from Hermitian Geometry. Thus, in this note, we give a rapid introduction to Kählerian/Hermitian geometry. (See Chapter 4 for details, where we essentially just go over the definitions of metrics, connections, curvatures and Laplacian.) Then we go through the Chern-Lu formulae in Chapter 5. As the conclusion of these efforts, the generalization of Schwarz's Lemma by Chern-Lu, for the holomorphic mappings from the complex n-dimensional open unit ball into a Hermitian manifold with holomorphic bisectional curvature bounded from above by a negative constant, is presented.

In 1970's and 1980's, further remarkable advancements occurred; the case of holomorphic mappings from a complete Kählerian manifold with its Ricci curvature tensor bounded from below by a (negative) constant into a Hermitian manifold with its holomorphic bisectional (or sectional) curvature bounded from above by a negative constant was elegantly treated by S.-T. Yau ([Yau 1978]) and H. Royden ([Royden 1980]). Their methods appeared to be quite different from each other when their papers first appeared. Yau used the Almost Maximum Principle, which is valid for complete Riemannian manifold, together with an ingenious (almost mysterious to many of us) choice of a function replacing the role of logarithmic function used before. That method also remedies the lack of Ahlfors' shrinking methods on the domain manifold. On the other hand, Royden made use of the special type of exhaustion functions of the domain manifold, relying upon the lower bound of the Ricci curvature of the Kähler manifold. Then Royden used methods that seemed to avoid the Almost Maximum Principle altogether, by exploiting the special exhaustion function and adjusting Ahlfors' "shrinking method" by means of this special exhaustion.

We therefore recapitulate the almost maximum principle (AMP) of Omori ([Omori 1967]) and Yau ([Yau 1975]) from the viewpoint of special exhaustion function. This was explained briefly in Chapter 6. Using this, one can re-illuminate the proof of generalizations by Yau and Royden. Thus we give Yau's proof in Chapter 7, explicating how Yau's choice of his auxiliary function replacing the role of traditional logarithmic function emerged. The main thrust of Royden's generalization of Schwarz's Lemma resides in that he treats the holomorphic mappings into Hermitian manifolds with negative holomorphic sectional curvature. That requires some more analysis on the curvature terms on top of the Chern-Lu formula type calculations, which we also explain here.

Of course research on generalization (or "variation" as Osserman put it in his article [Osserman 1999b]) of Schwarz's Lemma still continues. In Chapter 8, we list only a few related works that are more recent than the contents up to Chapter 7 of these notes.

There are other versions of generalized Schwarz's Lemma in different contexts (from those of these notes), that are especially useful for the study of Nevanlinna Theory of holomorphic curves and for Kobayashi-hyperbolicity problems. They involve more general bundles treated here. We feel that those theorems follow much the same philosophy as the topics here. However, we hope that these notes will make the reader interested in those directions also, and supply motivation for exploration of that interesting idea.

Then we should mention that there have been many recent papers on Schwarz's Lemma in various different viewpoints. When one considers holomorphic maps from non-Kählerian Hermitian manifolds; various types of assumptions on the torsion tensor have been imposed in those papers. It seems quite interesting to explore in that. However, we decided not to go into that realm with this writing. Just in case the reader takes interest in the developments in such a direction, we included some Hermitian geometry rudiments in Chapter 4, and some comments at the end of Chapter 8. On the other hand, we point out that such non-Kählerian consideration is not essential (in fact we never use them really) in this exposition which only deals with the holomorphic mappings from a Kähler manifold into a Hermitian manifolds, as the analysis involving the gradient and Laplacian takes place in the source manifold (that is Kählerian).

The bibliography section as well as citations in the main text of these notes are obviously far from being complete. This is solely due to the authors' shortcomings. Serious readers should look for themselves in the MathSciNet (TM) at the internet address `http://www.ams.org/mathscinet` for more reference items up to date.

We would like to thank colleagues and students who have read the draft and provided helpful comments. Our special thanks go to Ian Graham of Toronto (Canada) and Robert E. Greene of U.C.L.A. (U.S.) who read this manuscript and gave us their invaluable comments. Last, but not least, the first named author (Kim) would like to express his special thanks to the colleagues including, but not limited to, Kaushal Verma, Harish Seshadri, Gadadhar Misra and Gautam Bharali of The Indian Institue of Science in Bangalore for their hospitality during his visit in September 2008. Without their initiation and encouragements, this writing would not have been possible.

November 2009

k.t.k. & h.l.

Contents

Chapter 1

Some Fundamentals

The purpose of this chapter is to provide some basics which will be needed in Chapter 2. We review fundamentals such as the mean-value property, sub-mean-value property, various versions of the maximum principle, and other basic theorems that will be cited repeatedly in later part of these notes.

1.1 Mean-Value Property

The classical Schwarz's Lemma depends upon the maximum modulus principle for the modulus (i.e., the absolute value) of holomorphic functions. A function is said to be *holomorphic* if it is a continuously-differentiable (i.e., $\mathcal{C}^1$) complex-valued function defined on an open set, which satisfies the Cauchy-Riemann equation(s). Namely, if we denote by $f(z)$ a $\mathcal{C}^1$ function defined on an open subset Ω in the complex plane $\mathbb{C}$, and if we write $f(z) = u(x, y) + iv(x, y)$ where u and v are real-valued functions and $z = x + iy$, then f is *holomorphic* whenever it satisfies

$$\frac{\partial u}{\partial x} = \frac{\partial v}{\partial y}, \quad \frac{\partial u}{\partial y} = -\frac{\partial v}{\partial x}.$$

Then Green's theorem implies that, for any piecewise $\mathcal{C}^1$ curve Γ which bounds an open set, say Ω in the complex plane,

$$\int_{\Gamma} f(z) \, dz = 0$$

whenever f is holomorphic on Ω and is continuous on the closure of Ω, which is the same as the union $\Omega \cup \Gamma$. This is of course a special case of the well-known theorem of Cauchy.

An important consequence of this is the following

1

Theorem 1.1 (Cauchy's Integral Formula). *Let Ω be an open set in the complex plane $\mathbb{C}$, containing a region W and its boundary ∂W, where this boundary is a piecewise $\mathcal{C}^1$ curve oriented counterclockwise. If $f : \Omega \to \mathbb{C}$ is a holomorphic function, then*

$$f(z) = \frac{1}{2\pi i} \int_{\partial W} \frac{f(\zeta)}{\zeta - z} d\zeta$$

for every $z \in W$.

Then the maximum modulus principle follows by this Cauchy's integral formula. All these are well-known, but we shall briefly recall how the exposition goes. First consequence is the following averaging principle for holomorphic functions:

Theorem 1.2. *Let $f : \Omega \to \mathbb{C}$ be a holomorphic function on an open set $\Omega \subset \mathbb{C}$. If $r > 0$ and $z \in \Omega$ are given such that the closure $cl(D(z,r))$, of the open disc $(D(z,r))$ with radius r centered at z, is contained in Ω, then*

$$f(z) = \frac{1}{2\pi} \int_0^{2\pi} f(z + re^{it}) dt.$$

Note that, by separating the real and imaginary parts of f, the same formula holds for the real and imaginary part of f, respectively.

It is well-known that the real and imaginary parts of a holomorphic function here are harmonic functions. And conversely, harmonic functions are locally the real (or imaginary, respectively) part of a holomorphic function (cf., e.g., [Ahlfors 1966]). Thus the averaging principle above holds for harmonic functions defined on the plane.

Another important thrust of Cauchy's integral formula above is that every holomorphic function admits, locally, a power series development (i.e., the Taylor series). Hence harmonic functions, being locally the real part of a holomorphic function, also receive a real power series development. Real-valued functions with real variables that admit convergent power series developments are called *real-analytic*, and harmonic functions therefore are real-analytic. One feature of real-analyticity is the following unique continuation principle.

Theorem 1.3 (Unique Continuation). *If f and g are real-analytic, real-valued functions defined on a connected open subset Ω of the plane, and if the set $\{x \in \Omega \colon f(x) = g(x)\}$ contains a non-empty open subset, then f and g coincide on Ω.*

The proof is well-known: Let $Z = \{x \in \Omega \colon \text{All derivatives at } x \text{ of } f - g \text{ vanish}\}$. Then by continuity of f and g and their derivatives, Z is a closed subset of Ω. The existence of power series developments for $f - g$ implies that Z is open, as the set on which the power series development converges is open and any real analytic function vanishes if all the coefficients of the power series development vanish. Hence we must have that either $Z = \Omega$ or $Z = \emptyset$ (the empty set), as Ω is connected. Since Z contains the non-empty open subset of the set $\{x \in \Omega \colon f(x) = g(x)\}$ given in the hypothesis, we see that $Z = \Omega$. This completes the proof. $\qquad\square$

Of course it is well-known that the averaging principle and the real-analyticity of harmonic functions can be explicated without the help of complex analysis. We explain it briefly.

Definition 1.1. Let $u \colon U \to \mathbb{R}$ be a real-valued, twice differentiable function defined on an open set U in the plane $\mathbb{R}^2$. Such u is called *harmonic* if $\Delta u = 0$, where $\Delta = \dfrac{\partial^2}{\partial x^2} + \dfrac{\partial^2}{\partial y^2}$ denotes the standard Laplacian.

One of the main properties of a harmonic function is the following averaging principle. We shall set up some notation first: $\|x\|$ denotes the norm of $x \in \mathbb{R}^2$, namely the square root of the sum of squares of each component of x. We shall also use the standard notation for the line integral as in standard second-year calculus.

Theorem 1.4 (Mean-Value Property). *Let $u \colon U \to \mathbb{R}$ be a harmonic function defined on an open subset U of the plane $\mathbb{R}^2$, and let $p \in U$. Let $r > 0$ be such that the closed disc $cl(D)(p, r) := \{x \in \mathbb{R}^2 \colon \|x - p\| \le r\}$ is[1] contained in U. Then*

$$u(p) = \frac{1}{2\pi r} \int_{\partial D(p,r)} u \, ds,$$

and

$$u(p) = \frac{1}{\pi r^2} \iint_{D(p,r)} u \, dA,$$

where the line integral is over the boundary $\partial D(p, r)$ of the disc $D(p, r)$ oriented counterclockwise and, ds and dA represent the line element and the area element, respectively.

---

[1]We denote the closure of a set X by $cl(X)$. We do not use, in this note, the general topology notation $\overline{X}$ for the closure of X, because it may be confused with the complex conjugate.

Proof. Let $v(x) := u(rx + p)$ for $x \in \mathrm{cl}(D)(0,1)$. Then $v \colon \mathrm{cl}(D)(0,1) \to \mathbb{R}$ is harmonic. If ν denotes the outward unit normal vector to $S_r := \partial D(0, r)$ then, for any t with $0 < t \leq 1$, Stokes' theorem yields the following:

$$
\begin{aligned}
0 &= \iint_{D(0,t)} \Delta v \, dA = \int_{S_t} \nabla v \cdot \nu \, ds \\
&= \int_0^{2\pi} \nabla v(t\omega) \cdot \omega \, t d\theta, \qquad (\text{where } \omega = (\cos\theta, \sin\theta)) \\
&= t \int_0^{2\pi} \frac{\partial}{\partial t} \Big[v(t\omega) \Big] \, d\theta = t \frac{d}{dt} \int_0^{2\pi} v(t\omega) \, d\theta \\
&= t \frac{d}{dt} \Big\{ \frac{1}{t} \int_0^{2\pi} v(t\omega) \, t d\theta \Big\} \\
&= 2\pi t \frac{d}{dt} \Big\{ \frac{1}{2\pi t} \int_{S_t} v \, ds \Big\}.
\end{aligned}
$$

Hence the average integral $A(t) = \dfrac{1}{2\pi t} \displaystyle\int_{S_t} v \, ds$ is a constant function of t in the range $0 < t \leq 1$. The continuity of v implies that $\lim_{t \downarrow 0} A(t) = v(0)$. By definition of v, we obtain the first identity for u. The second identity is now an easy consequence of iterated integration. $\qquad\qquad\square$

This proof can be easily modified to give the following:

Theorem 1.5 (Sub-Mean-Value Property). *If $u \colon \Omega \to \mathbb{R}$ is a twice differentiable function defined on an open set Ω in $\mathbb{R}^2$ containing the closure of the disc $D(p, r)$ of radius r centered at p, and if $\Delta u \geq 0$ there, then*

$$
u(p) \leq \frac{1}{2\pi r} \int_{\partial D(p,r)} u \, ds.
$$

Moreover, the following holds

$$
u(p) \leq \frac{1}{\pi r^2} \iint_{D(p,r)} u \, dA.
$$

Construction of a detailed proof (which is really parallel to the proof of Theorem 1.4 given above) is left to the reader as an exercise.

The real-analyticity, in fact the existence of the complex Taylor series development of a holomorphic function f was a consequence of the analyticity of the Cauchy kernel $1/(\zeta - z)$ that appears in the Cauchy integral formula. In general, when such a type of reproducing formula holds for a function with real-analytic kernel, the function is also real-analytic. Thus the following theorem in particular implies the real-analyticity of harmonic functions "directly".

Theorem 1.6 (Poisson Integral Formula). *Let* $u \colon \Omega \to \mathbb{R}$ *be a harmonic function defined on an open set* Ω *in* $\mathbb{R}^2$ *containing the closure of the disc* $D(0, R)$ *of radius* R *centered at the origin* 0*. Then*

$$u(\zeta) = \frac{1}{2\pi} \int_0^{2\pi} \frac{R^2 - |\zeta|^2}{|Re^{it} - \zeta|^2} u(Re^{it}) dt$$

for any ζ *with* $|\zeta| < R$.

The function $P : \partial D(0, R) \times D(0, R) \to \mathbb{R}$ defined by

$$P(Z, \zeta) := \frac{|Z|^2 - |\zeta|^2}{|Z - \zeta|^2}$$

is called *the Poisson kernel function* for the disc $D(0, R)$.

1.2 Maximum Principle, I—Harmonic and Holomorphic Functions

We now present the maximum principle which will play an important role in establishing classical Schwarz's Lemma.

Theorem 1.7 (Maximum Principle). *If a harmonic function* $u \colon \Omega \to \mathbb{R}$ *defined in a domain (i.e., a connected open subset)* Ω *in* $\mathbb{R}^2$ *attains a local maximum, then* u *is a constant function.*

Proof. Suppose that u attains its local maximum for at p. Since Ω is open, there exists $r > 0$ such that the closed disc $\mathrm{cl}(D)(p, r)$ is contained in Ω. We now establish first that u is a constant function on $D(p, r)$.

Assume the contrary that u is not constant on the disc $D(p, r)$. Then there exists $q \in D(p, r)$ such that $u(p) > u(q)$. Let $\delta = u(p) - u(q)$. Then by continuity of u there exists ϵ with $0 < \epsilon << r$ such that $D(q, \epsilon) \subset \Omega$ and $u(p) > u(x) + \delta/2$ for every $x \in D(q, \epsilon)$. Let $\rho = \|p - q\|$. Then by the mean-value property (Theorem 1.4)

$$u(p) = \frac{1}{2\pi\rho} \int_{\partial D(p,\rho)} u \, ds$$

$$= \frac{1}{2\pi\rho} \left\{ \int_{\partial D(p,\rho) \cap D(q,\epsilon)} u \, ds + \int_{\partial D(p,\rho) \setminus D(q,\epsilon)} u \, ds \right\}$$

$$\leq \frac{1}{2\pi\rho} \left\{ (u(p) - \frac{\delta}{2})\ell_1 + u(p)\ell_2 \right\},$$

where ℓ_1 = the length of $(\partial D(p,\rho) \cap D(q,\epsilon))$ and ℓ_2 = the length of $(\partial D(p,\rho) \setminus D(q,\epsilon))$. Thus $\ell_1 + \ell_2 = 2\pi\rho$. This together with the above computation yields

$$u(p) \leq u(p) - \frac{\delta\ell_1}{4\pi\rho},$$

which is absurd. Therefore u has to be constant on the disc $D(p,r)$.

Finally, by the unique continuation principle for real-analytic functions (since harmonic functions are always real-analytic), it follows that u is constant on Ω. □

Now we turn our attention to holomorphic functions

Theorem 1.8 (Strong Maximum Modulus Principle). *Let $f : \Omega \to \mathbb{C}$ be a holomorphic function defined on a domain Ω in $\mathbb{C}$ into $\mathbb{C}$. If $|f|$ attains its local maximum at some point of Ω, then f is a constant function.*

Proof. If the local maximum were zero, then $|f|$ is identically zero, and consequently $f = 0$ in a small neighborhood of the local maximum point. f is then identically zero and consequently a constant function.

If the local maximum is positive, then notice that the function $\log|f(z)|$ is a real-valued harmonic function—easily verified by a direct differentiation—in a small (connected) open neighborhood of the local maximum point. Thus by the maximum principle for harmonic function (Theorem 1.7), $\log|f|$, and hence $|f|$ itself, is constant in the same neighborhood. Since $\log|f| = \frac{1}{2}\log(f\overline{f})$ is real analytic except where f vanishes, $\log|f|$ is constant on Ω. This leads us to conclude that the analytic function f is constant. □

Corollary 1.1 (Weak Maximum Modulus Principle). *Let $f : \Omega \to \mathbb{C}$ be a holomorphic function defined on a bounded domain Ω in $\mathbb{C}$, and let*

G be a sub-domain of Ω such that the closure $cl(G)$ of G is contained in Ω. Then

$$\max_{z \in cl(G)} |f(z)| = \max_{z \in \partial G} |f(z)|.$$

1.3 Maximum Principle, II—For Subharmonic Functions

The reader may skip this section for now, because for the exposition of these notes the contents of this section will be needed only a couple of times (Chapters 5 and 8).

General subharmonic functions are defined to be real-valued upper-semicontinuous functions satisfying the sub-mean-value property, the second inequality of Theorem 1.5—general subharmonic functions are even allowed to take $-\infty$ as values at some points (but not everywhere). However, in this section we shall consider only $\mathcal{C}^2$ subharmonic functions, say u, satisfying $\Delta u \geq 0$ on a domain Ω in the complex plane, because that is what we need almost all the time in these notes.

Therefore we present a version of maximum principle for ($\mathcal{C}^2$, or more generally, continuous) subharmonic functions; but it only concerns the global interior maximum. (This is due to the lack of unique continuation property.)

Theorem 1.9 (Maximum Principle for $\mathcal{C}^2$ Subharmonic Functions). *Let u be a $\mathcal{C}^2$ subharmonic function defined on a domain Ω in the complex plane $\mathbb{C}$. If u attains its maximum at a point, say p, in Ω, then u is constant.*

The proof is also well-known (see for instance [Gilbarg and Trudinger 1977], p. 15, Theorem 2.2): Recall that $u(p) = M := \max_\Omega u$. We now show that the set $Z := \{z \in \Omega : u(z) = M\}$ is open. For this purpose, let $z \in Z$. Choose $r > 0$ such that $cl(D(z,r)) \subset \Omega$. Then by sub-mean-value property we see that

$$0 = u(z) - M \leq \frac{1}{\pi r^2} \iint_{D(z,r)} (u(\zeta) - M) \, dA(\zeta) \leq 0.$$

This and the continuity of u imply that $u(\zeta) - M = 0$ for every $\zeta \in D(z,r)$. Hence z is an interior point of Z. Thus Z is open. On the other hand, Z is closed because u is continuous. Since Ω is closed and Z is non-empty, this implies that $Z = \Omega$.

Note therefore that such versions of the strong and weak maximum principles still work for the continuous subharmonic functions.

Corollary 1.2. *If Ω is a bounded domain in $\mathbb{C}$, and if $u : cl(\Omega) \to \mathbb{R}$ is a continuous function that is subharmonic on Ω, then*

$$\sup_{\Omega} u = \sup_{\partial\Omega} u.$$

However, the version of maximum principle for harmonic functions, in which the existence of the interior local maximum implies constancy of the function, fails for subharmonic functions. That has to do with the unique-continuation-principle that the harmonic functions satisfy, which property subharmonic functions do not enjoy. Here is an example (again, well-known): Let $h(t)$ be a $\mathcal{C}^2$ function satisfying

$$h(t) = \begin{cases} 0 & \text{if } 0 \le t \le 1/4 \\ t & \text{if } 1/2 \le t \le 1 \end{cases}$$

and $h''(t) \ge 0$ for every $t \in [0,1]$. Then the function $u(x,y) := h(\sqrt{x^2 + y^2})$ is a $\mathcal{C}^2$ convex function on the whole unit disc. Thus it is certainly a $\mathcal{C}^2$ subharmonic function there. Note that u does attain a local maximum at an interior point, say the origin, but it is clearly non-constant.

Chapter 2

Classical Schwarz's Lemma and the Poincaré Metric

In this chapter, we shall study the classical Schwarz's Lemma (from 1880, approximately) and Pick's generalization (in 1916) for the mappings from the open unit disc into itself. We also demonstrate a few applications and present the re-formulation of the Schwarz-Pick Lemma into a differential geometric form using the Poincaré metric of the disc.

2.1 Classical Schwarz's Lemma

The original form of classical Schwarz's Lemma is what the reader finds in almost any textbook on complex analysis:

Theorem 2.1 (Schwarz's Lemma). *Let $f : D \to D$ be a holomorphic map from the open unit disc $D = \{z \in \mathbb{C} \mid |z| < 1\}$ into itself. If $f(0) = 0$, then the following hold:*

(i) *$|f(z)| \leq |z|$ for every $z \in D$.*

(ii) *$|f'(0)| \leq 1$.*

(iii) *If the equality holds in (i) for some $z_0 \neq 0$, or if the equality holds in (ii), then $f(z) = cz$ for some constant $c \in \mathbb{C}$ with $|c| = 1$.*

Proof. Consider the function

$$g(z) = \begin{cases} \dfrac{f(z)}{z} & \text{if } z \neq 0 \\[2mm] f'(0) & \text{if } z = 0. \end{cases}$$

By a removable singularity theorem, this function is holomorphic on D.

Let r be an arbitrarily chosen constant with $0 < r < 1$. Define $g_r(z) := g(rz)$. Then by the Maximum Modulus Principle for holomorphic functions

9

it follows that

$$\max_{|z|\leq 1}|g_r(z)| = \max_{|z|=1}|g_r(z)| = \max_{|z|=1}\frac{|f(rz)|}{|rz|} = \frac{1}{r}\max_{|z|=1}|f(rz)| \leq \frac{1}{r}.$$

For every $z \in D$, we may let r converge to 1. So $|g(z)| \leq 1$ for every $z \in D$. Since $f(0) = 0$, one deduces immediately that $|f(z)| \leq |z|$ for every $z \in D$, which establishes (i).

For (ii), it suffices to re-read from what was just proved. Since $f'(0) = g(0)$, one gets $|f'(0)| \leq 1$.

For (iii), assume $|f'(0)| = 1$. Then $|g(0)| = 1$. By the Maximum Modulus Principle, $g(z)$ is then a constant function. But then this constant must have absolute value 1. This implies, with the definition of g, that $f(z) = cz$ for every $z \in D$ with $|c| = 1$.

Finally assume $|f(z_0)| = |z_0|$ for some $z_0 \in D \setminus \{0\}$. Then $|g(z_0)| = 1$. The argument we used just now again implies that $f(z) = cz$ for every $z \in D$ with $|c| = 1$. This ends the proof.   □

It is a well-known basic fact that Schwarz's Lemma above characterizes the biholomorphic self-maps, which we call *automorphisms* throughout this note, of the unit disc. In fact we present:

Theorem 2.2. *For the unit disc D in $\mathbb{C}$ the automorphism group $Aut\,D$ is given by*

$$Aut\,D = \left\{ z \mapsto e^{i\theta}\frac{z-\alpha}{1-\bar{\alpha}z} \; : \; \theta \in \mathbb{R}, \alpha \in D \right\}.$$

Proof. For each $\alpha \in D$, set $\varphi_a(z) = \dfrac{z+a}{1+\bar{a}z}$. Then it is easy to check that, whenever $a \in D$, $\varphi_a(D) \subset D$. Moreover, $\varphi_a \circ \varphi_{-a}(z) = z = \varphi_{-a} \circ \varphi_a(z)$ for every $z \in D$. Hence every $e^{i\theta}\dfrac{z-\alpha}{1-\bar{\alpha}z}$ is an element of Aut (D).

Conversely, let f be an arbitrary element in Aut (D). Then let $a = f^{-1}(0)$, i.e., $f(a) = 0$. With the notation above, define the map g by

$$g(\zeta) := f \circ \varphi_a(\zeta).$$

Then g is a holomorphic map sending the unit disc D into D, with the holomorphic inverse map $g^{-1} = \varphi_{-a} \circ f^{-1}$. By Schwarz's Lemma, since $g(0) = 0$ and $g^{-1}(0) = 0$, we have

$$|(f \circ \varphi_a)'(0)| \leq 1 \quad \text{and} \quad |(\varphi_{-a} \circ f^{-1})'(0)| \leq 1.$$

Direct calculation yields $\varphi_a'(a) = 1 - |a|^2$ and $\varphi_{-a}'(a) = (1 - |a|^2)^{-1}$. Consequently,

$$|f'(a)| \leq (1 - |a|^2)^{-1} \tag{2.1.1}$$

$$|(f^{-1})'(0)| \leq 1 - |a|^2. \tag{2.1.2}$$

On the other hand, we have an obvious identity: since $f \circ f^{-1}(\zeta) = \zeta$ for every $\zeta \in D$, we have $|f'(a)||(f^{-1})'(0)| = |(f \circ f^{-1})'(0)| = 1$. This together with (2.1.1) and (2.1.2) implies that the last four inequalities are in fact equalities. In particular $|(f \circ \varphi_a)'(0)| = 1$. Therefore there exists a real number θ such that $f \circ \varphi_a(\zeta) = e^{i\theta}\zeta$ for every $\zeta \in D$. Hence $f(\zeta) = e^{i\theta}\varphi_{-a}(\zeta)$, and the desired conclusion follows. $\qquad\square$

It is easily checked that the biholomorphic self-maps of any domain Ω in the complex plane $\mathbb{C}$ form a group under the law of composition; we denote it by Aut (Ω), and call it the *automorphism group* of Ω. It is instructive to verify directly that the composition of any two maps of the form given in Theorem 2.2 is again of that form, and to verify that the inverse of any one also has the same form.

2.2 Pick's Generalization

Now we present the following modification that appeared more than 35 years after the lemma above was first discovered:

Theorem 2.3 (Schwarz-Pick Lemma [Pick 1916]). *If $f : D \to D$ is a holomorphic function from the open unit disc D into itself, then*

$$\frac{|f'(z)|}{1 - |f(z)|^2} \leq \frac{1}{1 - |z|^2}$$

for every $z \in D$. Moreover, the equality holds at any point of D if and only if f is an automorphism of D.

Proof. Fix $z \in D$, and let ζ be the complex variable. Consider two automorphisms of D:

$$\varphi(\zeta) = \frac{\zeta + z}{1 + \bar{z}\zeta}, \quad \psi(\zeta) = \frac{\zeta - f(z)}{1 - \overline{f(z)}\zeta}.$$

Then the composition $F = \psi \circ f \circ \varphi$ maps the open unit disc D into itself with $F(0) = 0$. Therefore Schwarz's Lemma says that $|F'(0)| \leq 1$. Direct calculation verifies that it is equivalent to

$$\frac{|f'(z)|}{1 - |f(z)|^2} \leq \frac{1}{1 - |z|^2}.$$

The remaining claim then follows by (iii) of Schwarz's Lemma.   □

The reader who sees this modification for the first time might ask (naturally!) what its significance may be. One answer—which fits to the spirit of our exposition—is that it reveals the geometric nature of Schwarz's lemma, which is indeed the main theme of these lecture notes.[1]

Incidentally, we shall begin to change our viewpoint from here on with the concept of Hermitian metrics, emphasizing the differential geometric side of Schwarz's Lemma.

Before leaving this section we put a trivial comment: Schwarz's Lemma gives the estimate of the derivative by the original function, which is

$$|f'(z)| \leq \frac{1 - |f(z)|^2}{1 - |z|^2}.$$

Moreover, the lemma says that the maximum possible modulus of the derivative is achieved at some point if and only if f is a holomorphic automorphism of the unit disc D. (An exercise to the reader: *What happens to $f\colon D \to D$ when the equality holds?*)

2.3   The Poincaré Length and Distance

We now exploit a little bit of differential geometry. As the exposition progresses we shall need more and more contents from Differential Geometry, which we give a summary in the next chapter.

---

[1]On the other hand Pick himself seems to have been more interested in the *interpolation problem* which is: *Given the set of k points $z_1, \ldots, z_k$ and another set of points $w_1, \ldots, w_k$ in the unit disc D, does there exist a holomorphic map $f\colon D \to D$ such that $f(z_j) = w_j$ for each $j = 1, \ldots, k$?* The answer is the following, known as the Pick-Nevannlina interpolation theorem:

Theorem (Pick). *For any set of k points $z_1, \ldots, z_k$ and another set of points $w_1, \ldots, w_k$ in the unit disc D, there exist a holomorphic map $f\colon D \to D$ such that $f(z_j) = w_j$ for each $j = 1, \ldots, k$, if and only if the $k \times k$-matrix with (i,j)-th entry $\dfrac{1 - w_i \bar{w}_j}{1 - z_i \bar{z}_j}$ is positive definite.*

The *Poincaré metric* on the unit disc D is defined to be

$$ds_z^2 = \frac{dz \otimes d\bar{z}}{(1 - |z|^2)^2}.$$

This is a Hermitian inner product on the (holomorphic) tangent space $T_z D$. [For the concept of various tangent and co-tangent spaces for complex manifolds (as well as for domains in particular), see Chapter 4 of this note, especially Section 4.1.] We shall identify $T_z D$ with the complex plane $\mathbb{C}$. Then the preceding notation simply means

$$ds_z^2(v, w) = \langle v, w \rangle_z = \frac{v\bar{w}}{(1 - |z|^2)^2}.$$

The *pull-back* $f^* ds^2$ of ds^2 by the holomorphic map $f : D \to D$ is defined by

$$(f^* ds^2)_z(v, w) := ds_{f(z)}^2(df_z(v), df_z(w)).$$

With this concepts and notation, notice that the Schwarz-Pick Lemma says precisely the following:

Proposition 2.1. *If $f : D \to D$ is holomorphic, then $f^* ds^2 \le ds^2$.*

This of course implies

Corollary 2.1. *If $f \in Aut\ D$, then $f^* ds^2 = ds^2$.*

It is possible to translate it into expressions involving length of curves and the induced distance. We will do that before we progress further. Let $\gamma : [a, b] \to D$ be a $\mathcal{C}^1$ curve. Then the Poincaré length of γ is defined to be

$$L(\gamma) = \int_\gamma ds := \int_a^b ds_{\gamma(t)}(\gamma'(t)) dt.$$

Then the *Poincaré distance d* is defined in a customary way: $d(p, q)$ is defined to be the infimum of the lengths of the $\mathcal{C}^1$ curves in D joining p and q.

The Poincaré length of the curve γ is given explicitly by

$$L(\gamma) = \int_a^b \frac{|\gamma'(t)|}{1 - |\gamma(t)|^2} dt.$$

If a pair of points $p, q \in D$ have been given, and if we consider the γ's with $\gamma(a) = p$ and $\gamma(b) = q$, it is natural to ask whether there is a shortest connection from p to q. Writing expressions explicitly one obtains

$$L(\gamma) = \int_a^b \frac{\sqrt{(\text{Re } \gamma'(t))^2 + (\text{Im } \gamma'(t))^2}}{1 - (\text{Re } \gamma(t))^2 - (\text{Im } \gamma(t))^2} \, dt$$

$$\geq \int_a^b \frac{\text{Re } \gamma'(t)}{1 - (\text{Re } \gamma(t))^2} \, dt$$

$$\geq \tanh^{-1}(\text{Re } q) - \tanh^{-1}(\text{Re } p).$$

In the case when $p = 0 + 0i$ and $q = r + 0i$ with $0 < r < 1$, one obtains the conclusion from the preceding computation that the shortest connection between p and q is the straight line segment. Thus if we take the *Poincaré distance* $d(p,q)$ as earlier to be the infimum of all possible values of the Poincaré lengths of the curves joining 0 and $q > 0$, then $d(0,q) = \tanh^{-1} q$.

Due to Corollary 2.1 above, the Poincaré distance is invariant under the action of Möbius transforms on the unit disc. One sees then that the shortest connection between two points is the circular arc whose extension crosses the unit circle orthogonally, and that the distance formula is

$$d(p,q) = \tanh^{-1} \left| \frac{p - q}{1 - \bar{q}p} \right|.$$

The Schwarz-Pick Lemma implies that

$$d(f(p), f(q)) \leq d(p,q)$$

for any holomorphic map $f : D \to D$ and any points $p, q \in D$, since the Schwarz-Pick Lemma gives that f does not increase the Poincaré length of curves. In particular, for a holomorphic function $f : D \to D$ with $f(0) = 0$, the Schwarz-Pick Lemma implies that

$$\tanh^{-1} |f(z)| = d(f(0), f(z)) \leq d(0, z) = \tanh^{-1} |z|.$$

This is equivalent to (i) of the original Schwarz's Lemma.

Chapter 3

Ahlfors' Generalization

The previous chapter illustrated that fact that the Hermitian metric geometry is closely related to holomorphic mappings, at least for the unit disc. Indeed, it became a major theme in complex analysis that geometry and in particular curvature arose naturally in complex analysis. A specific form of this relationship is the principle

> *"Negative curvature restricts the behavior of holomorphic mappings."*

This principle in one form or another was often announced by Bochner, Chern and many others.

One of the initiating theorems in this line of thought is the theorem of Ahlfors ([Ahlfors 1938]) that we shall discuss now:

Let M be a Riemann surface, that is, a complex 1-dimensional complex manifold. Let ds_M^2 denote a Hermitian metric on M. Let ζ denote a local coordinate system. Then a Hermitian metric is represented by $ds_M^2 = h(\zeta)d\zeta \otimes d\bar{\zeta}$. The curvature is given by

$$K(\zeta) = -\frac{2}{h}\frac{\partial^2}{\partial\zeta\partial\bar{\zeta}}\log h,$$

a formula given by Gauss. A direct calculation verifies that the curvature of the Poincaré metric of the unit disc is -4.

3.1 Generalized Schwarz's Lemma by Ahlfors

Now we state and prove the generalization of Schwarz's Lemma by Ahlfors [Ahlfors 1938].

Theorem 3.1 (Ahlfors-Schwarz Lemma, 1938). *Let $f : D \to M$ be a holomorphic mapping. If M is a Riemann surface equipped with a Hermitian metric ds_M^2 with curvature bounded from above by a negative number $-K$, then*

$$f^* ds_M^2 \leq \frac{4}{K} ds_D^2$$

where ds_D^2 is the Poincaré metric of the unit disc D.

Proof. Since f is holomorphic, $f^* ds_M^2$ is a $(1,1)$-tensor on D. Thus,

$$f^* ds_M^2 = A(z)\, dz \otimes d\bar{z},$$

for some smooth function $A : D \to \mathbb{R}$.

Let $B(z) = \dfrac{1}{(1 - |z|^2)^2}$. It suffices to show that

$$\frac{A(z)}{B(z)} \leq \frac{4}{K}$$

for each $z \in D$. Following Ahlfors, we shall divide the proof into two cases.

<u>Special Case.</u> *Assume that the function $u(z) := A(z)/B(z)$ attains its maximum at z_0.*

Then of course it is enough to establish that

$$u(z_0) \leq \frac{4}{K}.$$

If the left-hand side is zero, there is nothing to prove. Hence we may assume that it is positive. Consequently in a small open neighborhood of z_0, the function u is positive.

Since the function u attains its maximum at z_0, we use the standard calculus to see that

$$\nabla \log u|_{z_0} = 0 \quad \text{and} \quad \Delta \log u|_{z_0} \leq 0,$$

where ∇ represent the gradient operator and Δ the Laplacian. A direct calculation with the conditions on the curvature then yields the estimate above. However, instead of leaving the details with the readers we shall briefly present the computation here.

At z_0, we have

$$0 \geq \Delta \log A - \Delta \log B.$$

Since the curvature of the Poincaré metric of unit disc is -4 we have

$$-\frac{1}{2B}\Delta \log B = -4.$$

From the upper bound condition of the curvature of ds_M^2 we also have

$$-\frac{1}{2A}\Delta \log A \le -K.$$

Now we obtain at z_0,

$$0 \ge 2AK - 8B.$$

Consequently,

$$u \le \frac{4}{K}.$$

However the assumption on the existence of maximum above does not hold in general. Thus we move to:

<u>General Case</u>. *Now $u : D \to \mathbb{R}$ is just non-negative and does not have to attain its maximum anywhere on D.*

Let $\xi \in D$ be arbitrarily chosen, and then fix it for a moment. We shall prove that $u(\xi) \le 4/K$.

Then consider a constant r with $|\xi| < r < 1$, and

$$D_r = \{z \in \mathbb{C} \mid |z| < r\}$$

and endow it with

$$ds_r^2 = B_r \, dz \otimes d\bar{z} = \frac{r^2 \, dz \otimes d\bar{z}}{(r^2 - |z|^2)^2}.$$

Let $f_r = f|_{D_r} : D_r \to M$. Then we see that $f_r^* ds_M^2 = u_r(z) \, ds_r^2$ with

$$u_r(z) = r^{-2}(r^2 - |z|^2)^2 A(z)$$

where A is a non-negative function on the whole disc D. Therefore, u_r is a non-negative function that vanishes on $\{z : |z| = r\}$. Therefore it attains its maximum on D_r.

Now one may apply the same calculation at the maximum point of u_r to obtain

$$f^* ds_M^2|_\xi \le \frac{4}{K} ds_r^2|_\xi.$$

Then, letting $r \to 1$ one gets the result. This completes the proof. $\qquad \square$

It may be worthwhile to re-appreciate this proof: First observe that both the source space and the target for the holomorphic mapping under consideration are complex 1-dimensional. This ensures that the pull-back of the Hermitian metric—a $(1,1)$-tensor—is a scalar function multiple of the metric of the source disc. Hence for the proof one is only to find the upper bound for the multiplier function in terms of curvatures. Use of Laplacian (as well as the gradient) at the maximum point is therefore entirely natural. The method of shrinking the disc that was used to remedy the general non-existence of maximum point is another important key point as mentioned several times. These lines of thoughts will appear repeatedly in subsequent developments.

3.2   Application to Kobayashi Hyperbolicity

In the geometric theory of holomorphic functions in several complex variables, the concept of invariant metric and distance plays an important role (cf., e.g., [Greene, Kim and Krantz 2010]). One of the primary examples is the Kobayashi distance and metric. Despite the terminology, these are only pseudo-distance or pseudo-metric in general—namely, the Hermitian property and the triangle inequality (for the distance, but not for the metric in general) hold but in general positive-definiteness does not. Therefore it is worth demonstrating that Ahlfors' generalization of Schwarz's Lemma gives a differential geometric criterion (in terms of curvature) for the positive-definiteness. We shall explain this aspect here, for the domains in $\mathbb{C}$ and Riemann surfaces only. But this continues to be valid in higher dimensions. (cf., [Kobayashi 1970], [Kobayashi 1998]).

As in Chapter 2, we continue using the notation d_D for the *Poincaré distance* for the open unit disc D in the complex plane $\mathbb{C}$. Denote by Hol (M, N) the set of holomorphic mappings from a Riemann surface M (or a domain in $\mathbb{C}$) into another such, say N. Define

$$\delta_M(p, q) = \inf\{d_D(a, b) \colon \exists \varphi \in \text{Hol } (D, M) \text{ such that}$$
$$\varphi(a) = p \text{ and } \varphi(b) = q \text{ for some } a, b \in D\}.$$

Now, by a *chain between p and q in M*, we mean a set of finitely many points $p_0, p_1, \ldots, p_N \in M$ satisfying $p - p_0$ and $p_N - q$. Then the *Kobayashi* *distance $d_M^K(p,q)$ between p and q in M* is defined to be

$$d_M^K(p,q) = \inf \sum_{j=0}^{N-1} \delta_M(p_j, p_{j+1})$$

where the infimum is taken over all possible chains between p and q in M.

Proposition 3.1 (Distance-Decreasing Property). *Let M, N be Riemann surfaces. If $f : M \to N$ is a holomorphic mapping, then $d_N^K\big(f(p), f(q)\big) \leq d_M^K(p,q)$ for any $p, q \in M$. In particular, if f is a biholomorphic mapping, then $d_N^K\big(f(p), f(q)\big) = d_M^K(p,q)$ for any $p, q \in M$.*

Since the proof follows by the definition and the property of the Poincaré metric, we shall not go through the detailed argument of the proof. However, it should be apparent to the reader that the Kobayashi distance can be an important concept for the study of holomorphic mappings in general.[1]

On the other hand, it is not at all clear whether the Kobayashi metric is positive-definite, i.e., whether $d_M^K(p,q) > 0$ whenever p and q are distinct points of M. It turns out that this property depends upon M, and in fact the Kobayashi distance is not always positive-definite. The reader can check quite easily that $d_{\mathbb{C}} = 0$:

Exercise: *Show that $d_{\mathbb{C}}(p,q) = 0$ for any $p, q \in \mathbb{C}$.* (Hint: Use Proposition 3.1 and maps $D \ni z \to Az + p \in \mathbb{C}$, and then let A diverge to ∞.)

So then, which condition will ensure positive-definiteness of the Kobayashi distance? In order to provide an answer via Schwarz's Lemma (Ahlfors' generalization), we shall exploit a well-known theorem of H.L. Royden in [Royden 1971].

On a complex manifold M (of course the Riemann surface case is included!) with the holomorphic tangent bundle $T'M$ (for the definition see Section 4.1 of these lecture notes), the *infinitesimal Kobayashi metric* (or, as it is often called the Kobayashi metric (or, the Kobayashi-Royden metric)) $k_M : T'M \to \mathbb{R}$ *of M* is defined to be

$$k_M(p; v) = \inf\{|\lambda| : \exists h \in \mathrm{Hol}\,(D, M) \text{ such that } h(0) = p, dh_0(\lambda) = v\}.$$

---

[1] The research concerning Kobayashi distance and metric became so extensive over decades; see [Kobayashi 1998] and references therein. Also we would like to make a remark on terminology: this distance-decreasing property is sometimes called *distance-non-increasing property* since distance can be at times preserved (by biholomorphic mappings for instance); but we choose to keep our choice as such, throughout this note.

Theorem 3.2 ([Royden 1971]). *The function $k_M \colon T'M \to \mathbb{R}$ is upper semi-continuous and,*

$$d_M^K(p,q) = \inf_\gamma \int_0^1 k_M(\gamma(t), \gamma'(t))\, dt$$

where the infimum is taken over all possible piecewise C^1 curve $\gamma \colon [0,1] \to M$ with $\gamma(0) = p$ and $\gamma(1) = q$.

We shall not provide the proof of this theorem here, as it is not the main stream of exposition of this lecture note. On the other hand, we shall now prove:

Proposition 3.2. *If M is a Riemann surface admitting a Hermitian metric with curvature bounded from above by -4, then the Kobayashi metric of M is positive-definite.*

Proof. The above-stated theorem of Royden tells us to establish a lower bound estimate for the infinitesimal Kobayashi metric k_M. Denote by $\| \ \|_p$ the Hermitian metric on M given in the statement of the Proposition. Let $f \colon D \to M$ be a holomorphic function from the unit disc D into M with $f(0) = p$ and $df_0(t) = v$. Then, by Ahlfors' generalization of Schwarz's Lemma (Theorem 3.1.1), we have

$$|t|^2 = \frac{|t|^2}{(1 - |0|^2)^2} \geq \|df_0(t)\|_p^2 = \|v\|_p^2,$$

The definition of the infinitesimal Kobayashi metric then implies

$$k_M(p,v)^2 \geq \|v\|_p^2,$$

as desired. $\square$

This proposition stays valid when M is a Hermitian manifold of arbitrary dimension. That will become obvious as the generalization of Schwarz's Lemma (which is the main theme of these notes) progresses along. On the other hand, the curvature bound does not have to be exactly -4; it can be any negative number, or even a negative function (For this last, cf. [Greene and Wu 1979]). Also, the Hermitian metric on M need not be complete in order for the proposition to be valid.

Following [Kobayashi 1967a], we call a complex manifold *hyperbolic* (or more precisely, *hyperbolic in the sense of Kobayashi*), if its Kobayashi distance is positive-definite. Call a complex manifold *complete hyperbolic* if its Kobayashi distance is complete in the sense that all Cauchy sequences converge.

One application of Kobayashi metric and the idea of hyperbolicity is as follows:

Proposition 3.3 ([Kobayashi 1970]). *If $f\colon \mathbb{C} \to M$ is a holomorphic mapping and if M is a Kobayashi hyperbolic complex manifold, then f is a constant mapping.*

Proof. Denote by d_M the Kobayashi distance of M. Then by the distance-decreasing property

$$d_M(f(z), f(0)) \le d_{\mathbb{C}}(z, 0) = 0$$

for any $z \in \mathbb{C}$. Since d_M is positive-definite, this yields that $f(z) = f(0)$ for any $z \in \mathbb{C}$. Hence f is constant. $\qquad\square$

Corollary 3.1 ([Kobayashi 1970]). *If M is a Riemann surface equipped with a metric with curvature bounded above by a negative constant, then every entire mapping from $\mathbb{C}$ into M is constant.*

It is worth mentioning that the complex plane minus two distinct points, which is of course biholomorphic to $\mathbb{C} \backslash \{0, 1\}$, admits a complete Hermitian metric with curvature ≤ -1. This result is due to H. Grauert and H. Reckziegel [Grauert and Reckziegel 1965] (See also pp. 12, Theorem 5.1, [Kobayashi 1970]). Therefore we see that the following famous theorem receives a geometric proof as an alternative to its original function-theoretic proof.

Theorem 3.3 (Little Picard Theorem). *Any entire function missing more than one point in its image is constant.*

Chapter 4

Fundamentals of Hermitian and Kählerian Geometry

We have arrived at a juncture where the Kählerian (a special case of Hermitian) differential geometry begins to be used extensively. So we now give a rapid introduction to complex differential geometry. A good reference for the reader (which is more extensive and comprehensive) is [Greene 1987]. Of course the classics [Chern 1979] and [Kobayashi and Nomizu 1969] are always highly recommended.

4.1 Almost Complex Structure

Let V be a vector space over the field $\mathbb{R}$ of real numbers. Assume that V admits a linear map $J : V \to V$ satisfying $J^2 = J \circ J = -I$ (where I represents the identity map). It is an exercise to show that $\dim V$ must be even in order for such a J to exist.

Such a J is called an *almost complex structure* on V and the vector space V equipped with J is called an *almost complex vector space*.

Now, consider[1] the complexification $V^{\mathbb{C}} := \mathbb{C} \otimes V$. The complex vector space $V^{\mathbb{C}}$ is of complex dimension $2m$. J extends to a complex linear map, with $J^2 = -I$.

The linear map J has only 2 eigenvalues $\pm i$. Consider the respective eigenspaces:

$$V' := \{v \in V^{\mathbb{C}} \mid Jv = iv\} \quad \text{and} \quad V'' := \{v \in V^{\mathbb{C}} \mid Jv = -iv\}.$$

Obviously, $V' \oplus V'' = V^{\mathbb{C}}$, and $\dim_{\mathbb{C}} V' = m = \dim_{\mathbb{C}} V''$. It is easy to

---

[1]The complexification can be understood as follows: if V has a basis $v_1, \ldots, v_N$. Then $V^{\mathbb{C}}$ is a linear span of $v_1, \ldots, v_N$ with complex number coefficients, where $v_1, \ldots, v_N$ are regarded linearly independent over the field of complex numbers.

verify that

$$V' = \{u - iJu \mid u \in V\} \quad \text{and} \quad V'' = \{u + iJu \mid u \in V\}.$$

4.2 Tangent Space and Bundle

Let M be a complex manifold of dimension m. Then it is also a smooth manifold. Let $p \in M$ and let T_pM be its tangent space, which is a vector space of dimension $2m$. Let TM denote the tangent bundle given by $TM = \bigcup_{p \in M} T_pM$, as usual in the manifold theory.

Since M is a complex manifold, it comes with the natural almost complex structure J, which we are going to describe now. We shall do it in terms of coordinates. Take a coordinate system $(z_1, \ldots, z_m) : U \to \mathbb{C}^m$ from a coordinate neighborhood U about $p \in M$. Write $z_k = x_k + iy_k$ for each k. Notice that the vectors

$$\frac{\partial}{\partial x_1}\Big|_p, \frac{\partial}{\partial y_1}\Big|_p, \quad \cdots \quad, \frac{\partial}{\partial x_m}\Big|_p, \frac{\partial}{\partial y_m}\Big|_p$$

span the real tangent space T_pM. Define $J_p : T_pM \to T_pM$ by

$$J_p\left(\frac{\partial}{\partial x_k}\Big|_p\right) = \frac{\partial}{\partial y_k}\Big|_p, J_p\left(\frac{\partial}{\partial y_k}\Big|_p\right) = -\frac{\partial}{\partial x_k}\Big|_p$$

for each $k = 1, 2, \ldots, m$ and extend it linearly over $\mathbb{R}$. Then $p \in M \mapsto J_p \in (T_pM)^* \otimes T_pM$ is a smooth map. Hence this correspondence shows that J is a smooth section of the bundle $T^*M \otimes TM$. This is an almost complex structure of M.

Now, we shall complexify T_pM, and consequently TM. We do this by extending coefficients. Namely, we let

$$\mathbb{C}T_pM := \mathbb{C} \otimes T_pM \quad \text{and} \quad \mathbb{C}TM := \mathbb{C} \otimes TM.$$

In local coordinates, the complexification simply means allowing complex values for coefficients for the real tangent vectors and tangent vector fields.

Extend J to the complex tangent spaces and bundles $\mathbb{C}$-linearly, following the formalism introduced above. Then consider the respective eigenspaces of J_p. They are

$$T_p'M = \{u - iJu \mid u \in T_pM\} \quad \text{and} \quad T_p''M = \{u + iJu \mid u \in T_pM\}$$

Traditional notation in local complex coordinates is worth mentioning at this juncture. They appear quite naturally now:

$$\frac{\partial}{\partial z_k} = \frac{1}{2}\left(\frac{\partial}{\partial x_k} - iJ\left(\frac{\partial}{\partial x_k}\right)\right) = \frac{1}{2}\left(\frac{\partial}{\partial x_k} - i\,\frac{\partial}{\partial y_k}\right)$$

and

$$\frac{\partial}{\partial \bar{z}_k} = \frac{1}{2}\left(\frac{\partial}{\partial x_k} + i\,\frac{\partial}{\partial y_k}\right),$$

where the factor $\frac{1}{2}$ is introduced for reasons one will soon see.

Notice that the Cauchy-Riemann equations for a mapping $f\colon M \to N$ between two complex manifolds M and N are equivalent to the equation $J_N \circ df = df \circ J_M$, where J_M, J_N are the almost complex structures constructed for M, N respectively.

One sees also that there is a natural $\mathbb{R}$-linear isomorphism (identification) between $T'_p M$ and $T_p M$ defined by

$$v \in T'_p M \mapsto \operatorname{Re} v \in T_p M.$$

Notice, however, that $T'_p M$ is a complex vector space of complex dimension m, whereas $T_p M$ is a real $2m$ dimensional space with no prescribed complex vector space structure.

Altogether, we have introduced four tangent spaces $T_p M, \mathbb{C}T_p M, T'_p M$ and $T''_p M$. They appear naturally for a complex manifold M, and of course they give rise to respective bundles.

4.3 Cotangent Space and Bundle

For the cotangent spaces and bundles, we shall simply build upon what we developed with the tangent spaces and bundles. The set of all $\mathbb{C}$-linear functionals on $\mathbb{C}T_p M$ will be the space we work in. With the basis

$$\frac{\partial}{\partial z_1},\dots,\frac{\partial}{\partial z_m}; \frac{\partial}{\partial \bar{z}_1},\dots,\frac{\partial}{\partial \bar{z}_m}$$

we shall take its dual basis. One can quickly check that the dual basis consist of complex co-vectors at p given by

$$dz_k := dx_k + idy_k, \qquad d\bar{z}_k := dx_k - idy_k,$$

for $k = 1,\dots,m$. (This is the reason for $\frac{1}{2}$ in the previous section because we customarily want $dz_j(\partial/\partial z_j) = 1$ and so forth.) Likewise one sees that

$T_p^{1,0}M := (T_p'M)^*$ is the vector space over $\mathbb{C}$ generated by $dz_1|_p, \ldots, dz_m|_p$, and that $T_p^{0,1}M := (T_p''M)^*$ by $d\bar{z}_1|_p, \ldots, d\bar{z}_m|_p$.

It may be a good practice for the sake of symbolic calculus, to verify the notational reasonability such as

$$df = \sum_{j=1}^m \frac{\partial f}{\partial z_j} dz_j + \sum_{j=1}^m \frac{\partial f}{\partial \bar{z}_j} d\bar{z}_j$$

for any smooth function $f : M \to \mathbb{C}$. Likewise one may define and develop the concept of complex differential forms of bi-degree (p, q) and their tensor products. However we shall not provide any further details.

4.3.1   *Hermitian metric*

We now introduce a Hermitian metric on a complex manifold M of complex dimension m. The passageway we take in this note is always through a real Riemannian geometry. Thus as usual, we restrict ourselves to the manifolds that are locally compact, Hausdorff, paracompact, second countable topological spaces.

Regard M as an almost complex manifold with the almost complex structure J introduced earlier. Then a *Hermitian metric* is a Riemannian metric h on M satisfying the condition

$$h_p(Jv, Jw) = h_p(v, w), \quad \forall v, w \in T_pM.$$

Then one may ask: *when can a complex manifold admit a Hermitian metric?* One always has a Rimannian metric, say g, thanks to the partitions of unity. The tensor $g(v, w) + g(Jv, Jw)$ then becomes immediately a Hermitian metric on M. Thus with our specifications on manifolds mentioned above, every complex manifold admits a Hermitian metric.

Recall that Hermitian metrics are defined on complex vector spaces and are complex-valued. There is a corresponding idea here. We start with a real-valued symmetric positive-definite Hermitian metric $h_p : T_pX \times T_pX \to \mathbb{R}$. Let $h_p' : T_p'X \times T_p'X \to \mathbb{C}$ be defined by

$$h_p'(v - iJv, w - iJw) = h_p(v, w) + i\, h_p(v, Jw),$$

for every $v, w \in T_pM$. The following are easy to check, and hence we leave the checking as an exercise for the reader:

(a) $h(v, Jw) = -h(w, Jv)$ for any $v, w \in T_pM$. Consequently, $h(v, Jv) = 0$ for any $v \in T_pM$.

(b) h' is Hermitian symmetric, i.e., $h'_p(V, W) = \overline{h'_p(W, V)}$ for any $V, W \in T'_pM$.

It is convenient for now to call h' the *complex Hermitian metric* corresponding to the real-valued Hermitian metric h.

4.4 Connection and Curvature

We now introduce the connections and curvatures briefly.

4.4.1 *Riemannian connection and curvature*

Let $\mathfrak{X}(M)$ denote the set of smooth vector fields on M.

Definition 4.1. A *linear connection* on the tangent bundle TM over the manifold M is a map $\nabla : \mathfrak{X}(M) \times \mathfrak{X}(M) \to \mathfrak{X}(M) : (X, Y) \mapsto \nabla_X Y$ satisfying:

(1) $\nabla_{f_1 X_1 + f_2 X_2} Y = f_1 \nabla_{X_1} Y + f_2 \nabla_{X_2} Y$ for any $f_1, f_2 \in \mathcal{C}^\infty(M)$ and any $X_1, X_2, Y \in \mathfrak{X}(M)$.

(2) $\nabla_X(aY_1 + bY_2) = a\nabla_X Y_1 + b\nabla_X Y_2$ for any $a, b \in \mathbb{R}$ and any $X, Y_1, Y_2 \in \mathfrak{X}(M)$.

(3) $\nabla_X(fY) = f\nabla_X Y + (Xf)Y$, for any $f \in \mathcal{C}^\infty(M)$ and any $X, Y \in \mathfrak{X}(M)$.

Linear connections are also called *affine connections*. For a differentiable manifold, there are infinitely many such connections. On the other hand, each such connection provides a method of differentiating a smooth vector field by another. Thus the linear connection is in fact a "differentiation".

Of course it is natural to look for a connection that can explain the particular geometry one aims to study. In our case that is the complex geometry, which concerns quantities such as the (almost) complex structure J and the Hermitian metric just introduced.

If we discount the complex structure concentrate on the metric structure (and consequently our manifold is just Riemannian), the natural and well-known connection is the *Levi-Civita connection* (i.e., the *Riemannian covariant differentiation*). Since the (real) Hermitian metric is Riemannian, we shall start with the Levi-Civita connection.

Definition 4.2. Let (M, h) be a Riemannian manifold. (The Hermitian metric h is also a real Riemannian metric.) Then the *Levi-Civita connection* on (M, h) is a linear connection ∇ satisfying the following two additional conditions:

(4) $\tau(X, Y) := \nabla_X Y - \nabla_Y X - [X, Y] = 0$

(5) $(\nabla h)(X, Y, Z) := X(h(Y, Z)) - h(\nabla_X Y, Z) - h(Y, \nabla_X Z) = 0,$

where the notation $[X, Y]$ stands for the Lie bracket of two vector fields X, Y.

It is well-known that the Levi-Civita connection exists and is unique (cf. [Greene 1987], [Kobayashi and Nomizu 1969], e.g.). The quantity τ is called the torsion tensor, and thus the (4) is called the torsion-free condition. (5) is commonly referred to as the condition that the metric is parallel. Of course this Levi-Civita connection is the key concept toward Riemannian geometry. It determines the geodesics, parallelism and the curvature.

4.4.2   *Riemann curvature tensor and sectional curvature*

Now we are ready to introduce the Riemannian curvature(s). In case the manifold is real two dimensional, the curvature is a function. However in higher dimensional case, the curvature is a multi-linear form on vector fields.

Let (M, J, h, ∇) be a complex manifold with a Hermitian metric h and its Levi-Civita connection ∇. We start with the (Riemannian) sectional curvature. Let $X, Y, Z, W \in \mathfrak{X}(M)$. Then we define the following notation:

$$R(X, Y)Z = \nabla_X \nabla_Y Z - \nabla_Y \nabla_X Z - \nabla_{[X,Y]} Z$$

$$R(X, Y, Z, W) = h(R(X, Y)Z, W).$$

Note that the last is a real-valued function, 4-linear on $\mathcal{C}^\infty(M)$. It is "pointwise" meaning that the value $R(X, Y, Z, W)\big|_p$ of $R(X, Y, Z, W)$ at $p \in M$ depends only on the point-values at p of the vector fields X, Y, Z and W.

Since this full curvature tensor is hard to use in general, one often considers the concept called the *Riemannian sectional curvature*. To define this, consider $X, Y \in \mathfrak{X}(M)$ that are linearly independent at $p \in M$ over $\mathbb{R}$. Then the value

$$K_p(X, Y) := -\frac{R(X, Y, X, Y)}{\|X \wedge Y\|^2}\bigg|_p$$

is the sectional curvature at p along the 2-dimensional plane in T_pM generated by X_p and Y_p, where $\|X \wedge Y\|^2 = h(X, X)h(Y, Y) - h(X, Y)^2$. It is not hard to check that this value of the sectional curvature depends only on the 2-dimensional plane (i.e., section) spanned by X_p and Y_p, but not on the choice of the basis vectors X_p and Y_p. In case the manifold is a real 2-dimensional surface in $\mathbb{R}^3$ equipped with the induced metric, that is its first fundamental form, then this sectional curvature coincides with the Gauss curvature.

4.4.3 *Holomorphic sectional curvature*

Now we re-instate the complex structure J back into consideration. Thus our manifold is now Hermitian. At this stage we have to re-consider our choice for the connection. Namely we have to consider which properties we would like to have for our linear connection to satisfy. Decision must be made among the following three properties:

(P1) $(\nabla h)(X, Y, Z) := X(h(Y, Z)) - h(\nabla_X Y, Z) - h(Y, \nabla_X Z) = 0$.
(P2) Torsion-free, i.e., $\tau(X, Y) := \nabla_X Y - \nabla_Y X - [X, Y] = 0$.
(P3) $(\nabla J)(X, Y) := \nabla_X(J(Y)) - J(\nabla_X Y) = 0$.

It is known that all three can be satisfied only if the metric h is special. Such a metric is called *Kählerian* (or simply *Kähler*). Several necessary and sufficient conditions for the metric to be Kähler are known as follows:

Proposition 4.1. *For a complex manifold M with the complex Hermitian metric h', consider a complex local coordinate system $(z_1, \ldots, z_n)$, and let $h'_{j\bar{k}} = h'\left(\dfrac{\partial}{\partial z_j}, \dfrac{\partial}{\partial z_j}\right)$ and $\omega = \sum h_{j\bar{k}} dz_j \wedge d\bar{z}_k$. Then the following are equivalent:*

(i) *h (or, equivalently, its complex form h') is Kähler, i.e., the Levi-Civita connection ∇ with respect to the metric h satisfies $\nabla J = 0$.*
(ii) *$d\omega = 0$.*

(iii) *There exists a smooth function φ such that $h'_{j\bar{k}} = \dfrac{\partial^2 \varphi}{\partial z_j \partial \bar{z}_k}$.*

Many well-known metrics are Kähler: the Poincaré metric on the disc and the Bergman metric of bounded domains in $\mathbb{C}^n$ are good examples.

On the other hand, general Hermitian metrics are not Kähler. In such a case what connection should be taken? It is generally agreed that condition (P3) $\nabla J = 0$ should be taken, but the "torsion-free" condition (P2) is dropped, allowing the *torsion tensor* τ in (P2) to be non-zero.

Regardless, when the manifold is Hermitian, one can make sense of "holomorphic sections"—those 2-dimensional plane in T_pM spanned by X_p and JX_p for some non-zero $X_p \in T_pM$ and the (Riemann) sectional curvature along this plane. Of course two vectors are linearly independent over $\mathbb{R}$ as we see from $h_p(X_p, JX_p) = 0$. Thus the *holomorphic sectional curvature* in the direction of X at p is defined to be $K_p(X, JX)$. (In Kählerian case, the holomorphic sectional curvature is indeed the Riemann sectional curvature for a holomorphic section.)

4.4.4 *The case of Poincaré metric of the unit disc*

We shall use the transitive automorphism group of the open unit disc to reconstruct the Poincaré metric. Let $G = \text{Aut } D$. Then consider the *isotropy subgroup* at the origin which is by definition $G_0 = \{g \in G \mid g(0) = 0\}$. From the explicit description of G, we know that G_0 consists of counterclockwise rotations.

So, on the tangent space T_0D the complex Euclidean Hermitian metric $dz \otimes d\bar{z}$ (or, its real part, if you prefer so) is invariant under the action of G_0. Now, for every $p \in D$, we shall describe the metric by

$$ds_p^2 = \mu^*(dz \otimes d\bar{z}),$$

where $\mu(z) = \dfrac{z - p}{1 - \bar{p}z}$. Hence the direct computation gives

$$ds_p^2 = \partial\mu|_p \otimes \overline{(\partial\mu|_p)} = \frac{dz|_p \otimes d\bar{z}|_p}{(1 - p\bar{p})^2}.$$

Hence the Poincaré metric, that is complex Hermitian, is

$$ds^2 = \frac{dz \otimes d\bar{z}}{(1 - z\bar{z})^2}.$$

Then we take the real part (for the real-valued Hermitian metric), which is (by an abuse of notation)

$$ds^2 = \frac{dx \otimes dx + dy \otimes dy}{(1 - x^2 - y^2)^2}$$

in the real (x, y) coordinates. (Here, $z = x + iy$, as usual.)

It is a good exercise to compute the Hermitian connection and the curvature, at least for a Riemann surface, following [Kobayashi and Nomizu 1969] for instance. In particular, the curvature is expressed in the following formula:

$$K(\zeta) = -\frac{2}{h} \frac{\partial^2}{\partial \zeta \partial \bar{\zeta}} \log h,$$

for the local expression of the Hermitian metric $ds^2 = h(\zeta) d\zeta \otimes d\bar{\zeta}$.

For the Poincaré metric the curvature is constant -4.

Remark 4.1. This Poincaré metric has higher dimensional version. The construction above yields the metric naturally, because the automorphism group of the open unit ball B^n of $\mathbb{C}^n$ is known to be generated by the unitary maps and the Möbius type maps

$$(z_1, \ldots, z_n) \mapsto \left(\frac{z_1 - \alpha}{1 - \bar{\alpha} z_1}, \frac{\sqrt{1 - |\alpha|^2}}{1 - \bar{\alpha} z_1} z_2, \ldots, \frac{\sqrt{1 - |\alpha|^2}}{1 - \bar{\alpha} z_1} z_n \right).$$

The Poincaré metric for the unit ball therefore turns out to be

$$ds^2 = \sum_{j,k=1}^{n} \left(\frac{\delta_{jk}}{(1 - \|z\|^2)} + \frac{\overline{z_j} z_k}{(1 - \|z\|^2)^2} \right) dz_j \otimes d\overline{z_k},$$

where δ_{jk} is the Kronecker delta, which is 1 if $j = k$ and 0 otherwise.

4.5 Connection and Curvature in Moving Frames

As is well-known, there are concepts called the *bisectional curvature* and the *Ricci curvature*. It is of course possible to introduce them by continuing the discussion of preceding section. However we choose not to do that. Instead, we are going to introduce Cartan's "moving frame method" that is more suitable for our purposes. We in particular use the moving frame method for the Chern-Lu formula in Chapter 5, as Chern (see Chapter 5) and Yau (see Chapter 7) did in their papers, respectively.

4.5.1 *Hermitian metric, frame and coframe*

Even though we deal mostly with Kählerian case (where the torsion tensor τ vanishes), it is going to be useful for the future developments to introduce the general Hermitian case.

Let $T'M$ represent the holomorphic tangent bundle. Given an Hermitian metric, it is possible to choose a smoothly varying orthonormal basis (usually called a *unitary frame*)

$$e_1, \ldots, e_m$$

in a local coordinate neighborhood. This can be done, for example, by applying the Hermitian Gram-Schmidt process to the coordinate frame $\dfrac{\partial}{\partial z_1}, \cdots, \dfrac{\partial}{\partial z_m}$. (Note that the unitary frame therefore is smooth, but not consisting of holomorphic vector fields in general.)

Then consider its dual, that is the (holomorphic) cotangent bundle $T^{1,0}M$, whose sections are called the (smooth) $(1,0)$-forms. Take the basis for sections of $T^{1,0}M$ dual to the frame chosen above and denote it by

$$\theta_1, \ldots, \theta_m.$$

This particular basis is called a *unitary coframe*.

Then the Hermitian metric can be written by

$$ds^2 = \sum_{i=1}^{m} \theta_i \otimes \bar{\theta}_i.$$

4.5.2   *Hermitian connection*

We now introduce the connection we shall use, continuing the discussion of the preceding section (with the same notation). We feel however that this part of exposition can be quite terse—thus we give an example here which illustrates how a connection can be interpreted in terms of a certain matrix of 1-forms. The reader may skip this example if they are familiar with such matters.

Example 4.1. Let (M, g) be a Riemannian manifold and let ∇ be the Levi-Civita connection. Take a local coordinate neighborhood and a local coordinate system $x_1, \ldots, x_m$. Let

$$e_j = \frac{\partial}{\partial x_j},$$

for $j = 1, \ldots, m$. Then it is customary to write

$$\nabla_{e_i} e_j = \sum_k \Gamma_{ij}^k e_k.$$

The functions Γ_{ij}^k are the (2nd) Christoffel symbols. The Leibniz rule which the connection ∇ satisfies is

$$\nabla_{e_i}(\psi e_j) = e_i(\psi) \cdot e_j + \psi \cdot \sum_k \Gamma_{ij}^k e_k.$$

Now, considering the meaning of the differential forms and the sections of bundles involved, one can now makes sense of the expression:

$$\nabla \colon \Gamma(TM) \to \Gamma(TM \otimes T^*M)$$

given by

$$\nabla\Big(\sum_{j=1}^{m} \psi^j e_j\Big) = \sum_{j=1}^{m} \Big((d\psi^j) \otimes e_j + \sum_{k=1}^{m} \psi^k \theta_{kj} \otimes e_j\Big).$$

The relation between the connection form (a matrix, in fact, of 1-forms) and the Levi-Civita connection ∇ should be visible from this, at least. (Of course this does not explain fully how all the other properties (such as torsion (free) condition, metric compatibity etc.) of connection matrix and related concepts (such as curvature and others) are developed and computed. For further information, cf., e.g., [Chern 1979] and [Chern 1968]).

We return to the Hermitian case and choose a suitable connection form on the m-dimensional Hermitian manifold M. Cartan's method says[2] that the connection matrix can be chosen from the following equation

$$d\theta_i = \sum_{j=1}^{m} \theta_j \wedge \theta_{ji} + \tau_i.$$

Notice that neither θ_{ji} nor τ_i are determined through this identity. Hence there are (infinitely) many choices for the connection form θ_{ji} and the torsion form τ_i. Rather, one needs to put extra assumptions in order to select the suitable connection matrix (as well as the torsion). A good example, which we use is the *canonical Hermitian connection* (i.e., the Chern connection), which is the choice of θ_{ji} satisfying the conditions:

$$\theta_{ij} + \overline{\theta_{ji}} = 0$$

and

$$\tau_i = \frac{1}{2} \sum_{j,k=1}^{m} T_{ijk}\theta_j \wedge \theta_k.$$

Note that this last requires that the torsion is of type $(2,0)$ only. (No $(1,1)$ part exists. And, of course, the whole τ vanishes in the Kähler case.)

---

[2] A good place the reader may find a comprehensive and yet concise introduction is [Chern 1989]; there he even claimed that this can be taught right after "vector calculus".

4.5.3   *Curvature*

The *curvature form* is defined to be

$$\Theta_{ij} = d\theta_{ij} - \sum_{k=1}^{m} \theta_{ik} \wedge \theta_{kj}.$$

One may check that the identity $\Theta_{ij} = -\overline{\Theta}_{ji}$ holds for the curvature form. Also,

$$\Theta_{ij} = \frac{1}{2} \sum_{k,\ell=1}^{m} R_{ijk\ell} \theta_k \wedge \overline{\theta}_\ell.$$

Namely, the curvature form Θ_{ij} are of type $(1,1)$. Notice that the skew-Hermitian symmetry for the curvature form above is equivalent to

$$R_{ijk\ell} = \overline{R_{ji\ell k}}.$$

In this notation, the holomorphic sectional curvature, the bisectional curvature and the Ricci curvature are easy to define. They are, respectively,

- The *holomorphic sectional curvature* in the direction of vector field $\eta = \sum_{k=1}^{m} \eta_k e_k$ is

$$\frac{\sum_{i,j,k,\ell=1}^{m} R_{ijk\ell} \eta_i \overline{\eta}_j \eta_k \overline{\eta}_\ell}{\left(\sum_{i=1}^{m} \eta_i \overline{\eta}_i\right)^2}.$$

- The *(holomorphic) bisectional curvature* determined by $\xi = \sum_{k=1}^{m} \xi_k e_k$ and $\eta = \sum_{k=1}^{m} \eta_k e_k$ is

$$\frac{\sum_{i,j,k,\ell=1}^{m} R_{ijk\ell} \xi_i \overline{\xi}_j \eta_k \overline{\eta}_\ell}{\left(\sum_{i=1}^{m} \xi_i \overline{\xi}_i\right)\left(\sum_{i=1}^{m} \eta_i \overline{\eta}_i\right)}.$$

- The *Ricci tensor* is given by

$$R_{ij} = \sum_{k=1}^{m} R_{ijkk},$$

and

$$\mathrm{Ric}(\xi, \eta) = \sum_{i,j=1}^{m} R_{ij} \xi_i \overline{\eta}_j.$$

4.5.4 The Hessian and the Laplacian

For a smooth function $u : M \to \mathbb{R}$ on the Riemannian manifold M, the *Hessian* of u is the *second covariant derivative* that is defined[3] to be, in the Riemannian covariant derivative notation,

$$Hess(u)(X,Y) = \nabla^2 u(X,Y) := X(Yu) - (\nabla_X Y)u$$

for every $X, Y \in \mathfrak{X}(M)$. The *Laplacian* Δu of u is defined as the *trace of Hess(u)*.

For Hermitian manifold M of real dimension $2m$, let $e_1, \ldots, e_{2m}$ be a real-orthonormal basis of $T_p M$. Then

$$\Delta u(p) = \sum_{i=1}^{2m} Hess\Big|_p (u)(e_i, e_i).$$

For the same Hermitian manifold M, the *complex Laplacian* of u, is defined using moving frame approach as follows: one writes

$$du = \sum_{i=1}^{m} u_i \theta_i + \sum_{i=1}^{m} \bar{u}_i \bar{\theta}_i.$$

Taking one more exterior derivative (with connection forms) one can define u'_{ij}, u_{ij} by

$$du_i - \sum_j u_j \theta_{ij} = \sum_j u'_{ij} \theta_j + u_{ij} \bar{\theta}_j.$$

Define the *complex Laplacian* of u by

$$\Delta_c u = \sum_i u_{ii}.$$

Remark 4.2. It is important to realize that the Laplacian of a function is the trace of its second covariant differentiation. Notice therefore that the Laplacian Δ_c above relies upon the canonical Hermitian connection ∇.

---

[3]The *trace of a bilinear form with respect to a given inner product* $g = \langle \cdot, \cdot \rangle$ is slightly different from the trace of a matrix. On a finite dimensional vector space V with an inner product, let $B : V \times V \to \mathbb{R}$ be a bilinear form. Then let $e_1, \ldots, e_m$ be an orthonormal basis. Then the trace of B with respect to the inner product given is defined to be

$$tr_g B = \sum_{j=1}^{m} B(e_j, e_j).$$

Notice that this definition is independent of the orthonormal basis. With respect to a general basis $v_1, \ldots, v_m$, the trace has a representation. Let $g_{ij} := \langle v_i, v_j \rangle$, and denote by g^{ij} the (i,j)-th entry of the inverse matrix of (g_{ij}). Also let $B_{ij} = B(v_i, v_j)$. Then it is known that $tr_g B = \sum_{i,j=1}^{m} g^{ij} B_{ij}$. The concept of trace in the Hermitian case is understood analogously.

Chapter 5

Chern-Lu Formulae

The further generalizations of Schwarz's Lemma by S.-S. Chern and Y.-C. Lu concern the holomorphic mappings $f : B^n \to M$ where B^n is the open unit ball in $\mathbb{C}^n$ and M a Kähler manifold of dimension m.

It is natural to recall the key ingredients of Ahlfors' method and establish a strategy:

First let us pull back the Hermitian metric, say h, of M by the holomorphic map f. Then f^*h is a $(1,1)$-tensor as is the Poincaré metric of B^n

$$ g = \sum_{j,k=1}^{n} \Big(\frac{\delta_{jk}}{(1 - \|z\|^2)} + \frac{\overline{z_j}z_k}{(1 - \|z\|^2)^2} \Big) dz_j \otimes d\overline{z_k}, $$

where δ_{jk} is equal to 1 if $j = k$, and 0 otherwise. (See Remark 4.1.) Schwarz's Lemma is concerned with the comparison of these g and f^*h.

On the other hand, since the complex dimension of the source manifold is not one, it is not obvious how to find a smooth, non-negative (also suitable and estimable) function $u : B^n \to \mathbb{R}$ satisfying

$$ f^*h \le ug. $$

Indeed the first important result of Chern-Lu analysis is that such a comparison function u exists.

Once such a u is found, one must go for an effective (upper-bound) estimate of u. As is done in the proof argument of Ahlfors' result one would like to apply the maximum principle.

First assume the special case when u attains its maximum, say at a point $p \in B^n$. If $u(p) = 0$, there is nothing to prove. So assume that $u(p) > 0$. Then, at p, we shall see that the gradient $\nabla \log u$ is equal to 0 and the Laplacian $\Delta \log u$ is non-positive. However one must realize that the gradient and the Laplacian here are based upon the Riemannian metric and its Levi-Civita connection.

In the light of Ahlfors' arguments, one expect for the relations $\nabla \log u|_p = 0$ and $\Delta \log u|_p \leq 0$ to yield an (efficient) upper-bound estimate for u by the ratio of the curvature bounds. This will require a suitable formula for the gradient and Laplacian of log of the comparison function. This is the key result of this chapter: the Chern-Lu formulae.

On the other hand, it is worth noting that the Chern-Lu formulae uses the Laplacian based upon the Hermitian connection! That is why the use of Chern-Lu formula for general Schwarz's lemmas can be effective only for the holomorphic mappings from a Kählerian manifold; the Laplacians, one coming from the Levi-Civita connection and the other from the Hermitian connection, coincide except for the constant multiplier $1/2$ which is immaterial.

In the general case, the comparison function u may not attain its maximum. In order to remedy the non-existence of maximum points, one may imitate Ahlfors' shrinking technique if possible. This is why Chern as well as Lu requires the domain manifold to be the ball.

Now we have explained how the strategy towards this version of general Schwarz's lemma is set up. We shall present the details following the papers of Chern and Lu ([Chern 1968], [Lu 1968]).

One final remark before beginning to introduce the analysis of Chern and Lu in the next section: there had been earlier investigations of high dimensional case similar to Chern-Lu result (cf., e.g., [Kobayashi 1967a]).

5.1 Pull-Back Metric against the Original

Consider a very general setting: Let (M, g) and (N, h) be Hermitian manifolds of complex dimension m and n respectively. Let $f : M \to N$ be a holomorphic mapping. The goal of this section is to compare f^*h and g on M.

Let us first arrange the indices. The roman indices $i, j, k, \ldots$ will run from 1 through $m = \dim M$, and the Greek $\alpha, \beta, \ldots$ from 1 through $n = \dim N$.

Denote by $\theta_1, \ldots, \theta_m$ a coframe for M, and by ω_α the same for N. Then

$$f^*\omega_\alpha = \sum_{i=1}^m a_{\alpha i}\theta_i$$

for some smooth real-valued functions $a_{\alpha i}$. Thus

$$f^*(h) = f^*(\omega_\alpha \otimes \overline{\omega}_\alpha) = \sum_{\alpha,i,j} a_{\alpha i} \overline{a_{\alpha j}} \, \theta_i \otimes \overline{\theta_j}.$$

Define by

$$(b_{ij}) = \left(\sum_\alpha a_{\alpha i} \overline{a_{\alpha j}} \right).$$

Note that the matrix (b_{ij}) is Hermitian symmetric and positive semi-definite.

Let λ_i be the eigenvalues of the matrix (b_{ij}). Then Linear Algebra tells us that there exists a certain (unitary) coframe ϑ_i such that

$$f^*h = \sum_i \lambda_i \, \vartheta_i \otimes \overline{\vartheta}_i$$

$$\leq \sum_i \lambda_i \sum_i \vartheta_i \otimes \overline{\vartheta}_i.$$

Hence it is appropriate to let

$$u := \sum_i \lambda_i = \operatorname{tr}\,(b_{ij}) = \sum_{\alpha,i} a_{\alpha i} \overline{a_{\alpha i}}.$$

Then one has that

$$f^*(h) \leq u \, g,$$

which is, in effect, the most natural way to compare f^*h to g. (A priori, one might look at $\max |\lambda_i|$ or the like. But it turns out that trace is the easiest to deal with.)

5.2 Connection, Curvature and Laplacian

In order to apply Ahlfors' method to u, one needs to look at the fundamentals such as connection, curvature and Laplacian. (See Chapter 4 for general summary.)

Start with the structure equation with unitary coframe θ_i for a Hermitian manifold (M, g). Since the exterior derivative $d\theta_i$ is a 2-form, one may choose 1-forms θ_{ij} and 2-forms Θ_i satisfying

$$d\theta_i = \sum_j \theta_j \wedge \theta_{ji} + \Theta_i.$$

Of course there is no reason at this stage that the choices for θ_{ji} and Θ_i have to be unique. But we have mentioned in Chapter 4 that one can further require that the following conditions are met:

$$\theta_{ij} + \bar{\theta}_{ji} = 0,$$

and

$$\Theta_i = \frac{1}{2} \sum_{j,k} T_{ijk} \theta_j \wedge \theta_k,$$

for some smooth functions T_{ijk}. (This last requires that Θ_i's are $(2,0)$-forms.)

The matrix (θ_{ij}) with 1-forms as its entries is called the *connection matrix*. The 2-forms Θ_i's are called the *torsion*. Take exterior derivative of structure equation of $d\theta_i$ to obtain

$$d\Theta_i = \sum_j \theta_j \wedge \Theta_{ji} - \sum_j \Theta_j \wedge \theta_{ji},$$

where

$$\Theta_{ji} = d\theta_{ji} - \sum_k \theta_{jk} \wedge \theta_{ki}.$$

This is actually a $(1,1)$-form satisfying

$$\Theta_{ij} + \overline{\Theta_{ji}} = 0.$$

The $(1,1)$-form Θ_{ij} can be written as

$$\Theta_{ij} = \frac{1}{2} \sum_{k,\ell} R_{ijk\ell} \theta_k \wedge \bar{\theta}_\ell,$$

where $R_{ijk\ell}$ is called the (coefficients of the) *curvature tensor*.

Now we clarify the notation again. For (M, g), we list the forms in structure equation as follows:

$$\theta_i, \ \theta_{ij}, \ \Theta_i, \ \Theta_{ij}, \ R_{ijk\ell}.$$

Likewise for (N, h), we list corresponding forms:

$$\omega_\alpha, \ \omega_{\alpha\beta}, \ \Omega_\alpha, \ \Omega_{\alpha\beta}, \ S_{\alpha\beta\gamma\eta}.$$

It is time to introduce the Laplacian for the smooth $(\mathcal{C}^\infty)$ real-valued function u on M. Although we are primarily interested in the function

u constructed above as the trace of the non-negative Hermitian matrix, the concept of the Laplacian introduced here is applicable for any general function u. From here on therefore, u can be regarded as an arbitrary smooth real-valued function on M (of course, $u > 0$ when we discuss $\log u$). Now, we shall begin with introducing the second covariant derivative of u using structure equation involving θ_i.

$$du = \sum_i u_i \theta_i + \sum_i \bar{u}_i \bar{\theta}_i.$$

Taking its exterior derivative and using structure equation of $d\theta_i$, one has

$$\sum_i (du_i - \sum_j u_j \theta_{ij}) \wedge \theta_i + \sum_i (d\bar{u}_i - \sum_j \bar{u}_j \bar{\theta}_{ij}) \wedge \bar{\theta}_i$$
$$+ \sum_i u_i \Theta_i + \sum_i \bar{u}_i \bar{\Theta}_i = 0.$$

Let

$$du_i - \sum_j u_j \theta_{ij} = \sum_j (u'_{ij} \theta_j + u_{ij} \bar{\theta}_j).$$

Applying it to the previous equation and separating the forms by their types, one arrives at

$$\sum_{i,j} u'_{ij} \, \theta_j \wedge \theta_i + \sum_i u_i \Theta_i = 0.$$

Thus one obtains

$$\sum_i d(u_i \theta_i) = \sum_i (du_i - \sum_j u_j \theta_{ij}) \wedge \theta_i + \sum_i u_i \Theta_i$$
$$= -\sum_{i,j} u_{ij} \, \theta_i \wedge \bar{\theta}_j.$$

The *complex Laplacian* of u is defined to be

$$\Delta_c \, u = \sum_i u_{ii}.$$

If $u > 0$, the following formula for $\log u$ turns out to be useful:

$$\Delta_c \log u = \frac{1}{u} \Delta_c u - \frac{1}{u^2} \sum_i u_i \bar{u}_i.$$

5.3 Chern-Lu Formulae

Our present goal is to compute $\Delta_c u$ and $\Delta_c \log u$ so that they are represented by the curvature terms of M and N via f.

Recall that θ_i, ω_α are coframe fields of (M, g) and (N, h), respectively. Let $f : M \to N$ be a holomorphic mapping such that

$$f^* \omega_\alpha = \sum_{i=1}^m a_{\alpha i} \theta_i,$$

or we shall use the following short-hand notation:

$$\omega_\alpha = \sum_i a_{\alpha i} \theta_i.$$

Of course we keep in mind that ω_α is indeed $f^* \omega_\alpha$ in what follows.

The first stage of computation involves obtaining proper expressions of the first and second covariant derivatives of $a_{\alpha i}$ through the exterior derivatives of the pull-back of the coframe ω_α of N. Notice that

$$d\omega_\alpha = \sum_i (da_{\alpha i} \wedge \theta_i + a_{\alpha i} d\theta_i).$$

Using the structure equation of θ_i and ω_α we have

$$\sum_\beta \omega_\beta \wedge \omega_{\beta\alpha} + \Omega_\alpha = \sum_i (da_{\alpha i} \wedge \theta_i + a_{\alpha i} \Theta_i) + \sum_{i,j} a_{\alpha i} \theta_j \wedge \theta_{ji}.$$

Taking θ_i as the common factor for a few terms, one obtains

$$\sum_i \left(da_{\alpha i} - \sum_j a_{\alpha j} \theta_{ij} + \sum_\beta a_{\beta i} \omega_{\beta\alpha} \right) \wedge \theta_i + \sum_i a_{\alpha i} \Theta_i - \Omega_\alpha = 0.$$

Since the torsion terms are of bidegree $(2, 0)$, we put

$$da_{\alpha i} - \sum_j a_{\alpha j} \theta_{ij} + \sum_\beta a_{\beta i} \omega_{\beta\alpha} = \sum_k{}' a_{\alpha i k} \theta_k.$$

Take its exterior derivative again to obtain an expression of $da_{\alpha i k}$ to obtain

$$-\sum_j da_{\alpha j} \wedge \theta_{ij} - \sum_j a_{\alpha j} d\theta_{ij} + \sum_\beta da_{\beta i} \wedge \omega_{\beta\alpha} + \sum_\beta a_{\beta i} d\omega_{\beta\alpha}$$

$$= \sum_k da_{\alpha i k} \wedge \theta_k + \sum_k a_{\alpha i k} d\theta_k.$$

The first and third terms of the left-hand side can be reformulated using the first covariant derivative formula of $a_{\alpha i}$. For the other terms, we use structure equation of θ_i, ω_α again. Then one can re-organize the preceding identity as follows:

$$\sum_k \left(da_{\alpha ik} - \sum_j a_{\alpha ij}\theta_{kj} - \sum_j a_{\alpha jk}\theta_{ij} + \sum_\beta a_{\beta ik}\omega_{\beta\alpha} \right) \wedge \theta_k$$

$$= -\sum_k a_{\alpha ik}\Theta_k - \sum_j a_{\alpha j}\Theta_{ij} + \sum_\beta a_{\beta i}\Omega_{\beta\alpha}.$$

Since Θ_k is a $(2,0)$-form, one may let

$$\sum_k a_{\alpha ik}\Theta_k = \sum_{k,\ell} a_{\alpha ik\ell}\theta_k \wedge \theta_\ell.$$

Using the defining equation of Θ_{ij} and

$$\Omega_{\beta\alpha} = \frac{1}{2}\sum_{\gamma,\eta} S_{\beta\alpha\gamma\eta}\,\omega_\gamma \wedge \bar\omega_\eta$$

$$= \frac{1}{2}\sum_{i,j}\sum_{\gamma,\eta} S_{\beta\alpha\gamma\eta}a_{\gamma i}\bar a_{\eta j}\,\theta_i \wedge \bar\theta_j$$

we set

$$\sum_j a_{\alpha j}\Theta_{ij} - \sum_\beta a_{\beta i}\Omega_{\beta\alpha} = \sum_{k,\ell} b_{\alpha ik\ell}\theta_k \wedge \bar\theta_\ell,$$

where

$$b_{\alpha ik\ell} = \frac{1}{2}\left(\sum_j a_{\alpha j}R_{ijk\ell} - \sum_{\beta,\gamma,\eta} a_{\beta i}a_{\gamma k}\bar a_{\eta\ell}\,S_{\beta\alpha\gamma\eta} \right).$$

Then one obtains

$$da_{\alpha ik} - \sum_j a_{\alpha ij}\,\theta_{kj} - \sum_j a_{\alpha jk}\,\theta_{ij} + \sum_\beta a_{\beta ik}\,\omega_{\beta\alpha} = \sum_\ell a_{\alpha ik\ell}\,\theta_\ell + \sum_\ell b_{\alpha ik\ell}\,\bar\theta_\ell.$$

Recall that the Ricci tensor is defined to be

$$R_{ij} = \sum_k R_{ijkk}.$$

We are now ready to state the Chern-Lu formula.

Theorem 5.1 (Chern-Lu Formulae). *With the settings above, one has*

$$\Delta_c\, u = \sum_{\alpha,i,k} |a_{\alpha ik}|^2 + \frac{1}{2}\sum_{\alpha,i,j} a_{\alpha i}\bar a_{\alpha j}R_{ij} - \frac{1}{2}\sum_{i,j}\sum_{\alpha,\beta,\gamma,\eta} a_{\alpha i}\bar a_{\beta i}a_{\gamma j}\bar a_{\eta j}S_{\alpha\beta\gamma\eta}$$

and

$$\Delta_c \log u = \frac{1}{2u}\left(\sum_{\alpha,i,j} a_{\alpha i}\bar a_{\alpha j}R_{ij} - \sum_{i,j}\sum_{\alpha,\beta,\gamma,\eta} a_{\alpha i}\bar a_{\beta i}a_{\gamma j}\bar a_{\eta j}S_{\alpha\beta\gamma\eta} \right).$$

Proof. Recall that

$$du = \sum_j u_j \theta_j + \sum_j \bar{u}_j \bar{\theta}_j$$

$$\sum_j d(u_j \theta_j) = -\sum_{j,k} u_{jk}\, \theta_j \wedge \bar{\theta}_k$$

$$\Delta_c u = \sum_j u_{jj}.$$

Here, u is defined by the pull-back of the metric as above:

$$u = \sum_{\alpha,i} a_{\alpha i} \bar{a}_{\alpha i}.$$

Take its exterior derivative and then extract the coefficients of θ_j and $\bar{\theta}_j$. A direct calculation yields

$$du = \sum_{\alpha,i} \bar{a}_{\alpha i} da_{\alpha i} + a_{\alpha i} d\bar{a}_{\alpha i}$$

$$= \sum_{\alpha,i} \bar{a}_{\alpha i} \Big(\sum_j a_{\alpha i j}\theta_j + \sum_j a_{\alpha j}\theta_{ij} - \sum_\beta a_{\beta i}\omega_{\beta\alpha} \Big)$$

$$+ \sum_{\alpha,i} a_{\alpha i} \Big(\sum_j \bar{a}_{\alpha i j}\bar{\theta}_j + \sum_j \bar{a}_{\alpha j}\bar{\theta}_{ij} - \sum_\beta \bar{a}_{\beta i}\bar{\omega}_{\beta\alpha} \Big).$$

Since

$$\theta_{ij} + \bar{\theta}_{ji} = 0 \qquad \omega_{\alpha\beta} + \bar{\omega}_{\beta\alpha} = 0,$$

it follows that

$$du = \sum_{\alpha,i,j} \bar{a}_{\alpha i} a_{\alpha i j}\theta_j + \sum_{\alpha,i,j} a_{\alpha i} \bar{a}_{\alpha i j}\bar{\theta}_j.$$

Thus

$$u_j = \sum_{\alpha,i} \bar{a}_{\alpha i} a_{\alpha i j}.$$

Now we compute u_{jk} by taking exterior derivative of u_j and finding the coefficients of $\bar{\theta}_k$. One has

$$du_j = \sum_{\alpha,i} \bar{a}_{\alpha i} da_{\alpha i j} + a_{\alpha i j} d\bar{a}_{\alpha i}.$$

Using the first and second covariant derivatives for $a_{\alpha i}$, which actually define the terms $a_{\alpha i k}$ and $a_{\alpha i k \ell}$, one sees that

$$du_j - \sum_k u_k \theta_{jk} = \sum_{\alpha,i} \bar{a}_{\alpha i} \Big(\sum_k a_{\alpha i k} \theta_{jk} + \sum_k a_{\alpha k j} \theta_{ik} - \sum_\beta a_{\beta i j} \omega_{\beta \alpha} $$
$$+ \sum_\ell a_{\alpha i j \ell} \theta_\ell + \sum_\ell b_{\alpha i j \ell} \bar{\theta}_\ell \Big)$$
$$+ \sum_{\alpha,i} a_{\alpha i j} \left(\sum_j \bar{a}_{\alpha j} \bar{\theta}_{ij} - \sum_\beta \bar{a}_{\beta i} \bar{\omega}_{\beta \alpha} + \sum_k \bar{a}_{\alpha i k} \bar{\theta}_k \right)$$
$$- \sum_k \sum_{\alpha,i} \bar{a}_{\alpha i} a_{\alpha i k} \theta_{jk}.$$

Identifying the coefficients of $\bar{\theta}_k$, one finds

$$u_{jk} = \sum_{\alpha,i} \bar{a}_{\alpha i} b_{\alpha i j k} + a_{\alpha i j} \bar{a}_{\alpha i k}.$$

Therefore the complex Laplacian of u is as follows:

$$\sum_k u_{kk} = \sum_{\alpha,i,k} |a_{\alpha i k}|^2 + \sum_{\alpha,i,k} \bar{a}_{\alpha i} b_{\alpha i k k}$$
$$= \sum_{\alpha,i,k} |a_{\alpha i k}|^2 + \frac{1}{2} \sum_{\alpha,i,j,k} a_{\alpha i} \bar{a}_{\alpha j} R_{ijkk} - \frac{1}{2} \sum_{i,j,\alpha,\beta,\gamma,\eta} a_{\alpha i} \bar{a}_{\beta i} a_{\gamma j} \bar{a}_{\eta j} S_{\alpha \beta \gamma \eta},$$

which yields the first formula in the assertion. From

$$u \sum_{\alpha,i,k} |a_{\alpha i k}|^2 - \sum_j u_j \bar{u}_j = 0$$

the second formula follows. This completes the proof. □

5.4 General Schwarz's Lemma by Chern-Lu

The Chern-Lu generalization of Schwarz's Lemma is as follows:

Theorem 5.2 (Chern/Lu, 1968). *Let B^n be the open unit ball in $\mathbb{C}^n$ equipped with the Poincaré-Bergman metric g with its Ricci curvature equal to the negative constant $-2n(n+1)$. Let (M, h) be a Kählerian manifold of complex dimension n with its holomorphic bisectional curvature bounded above by $-2n(n+1)$. Then, for every holomorphic mapping $f : B^n \to M$, the inequality*

$$f^* h \le g$$

holds.

Since the curvature bounds depend directly on multiplications of positive constants to the metrics, they can be regulated easily as long as the signs are unchanged. Hence we do not concern ourselves with the constants here. The reader should pay more attention to the role of the Chern-Lu formula in the proof.

A rough sketch of the proof. The essential step of the proof is the Chern-Lu formulae (see Theorem 5.1; also the contents of Section 5.1 for the definition of function u). And the remaining argument, which we present here, is just a straightforward modification of Ahlfors' argument presented in Chapter 3.

By the argument of Section 5.1, we have

$$f^*h \leq u \cdot g,$$

where u in particular is defined to be a smooth function on the unit ball B^n. Hence we first work with

Special case: u *attains its maximum at some point* $p \in B^n$.

If $u(p) = 0$, then there is nothing to argue. Hence we may assume without loss of generality that $u(p) > 0$. Then of course, at p, we can take a local coordinate neighborhood and see that

$$\nabla_c \log u = 0 \text{ and } \Delta_c \log u \leq 0.$$

(As remarked earlier, these hold for Riemannian gradient and Laplacian. But, since the Poincaré metric is Kählerian, the above result holds because the gradients and Laplacians coincide respectively up to constant multipliers.) By Theorem 5.1, this implies that

$$0 \geq \Delta_c \log u - \frac{1}{2u} \left(\sum_{\alpha,i,j} u_{\alpha i} \bar{u}_{\alpha j} R_{ij} - \sum_{i,j} \sum_{\alpha,\beta,\gamma,\eta} \bar{a}_{\alpha i} \bar{a}_{\beta i} \bar{a}_{\gamma j} a_{\eta j} S'_{\alpha\beta\gamma\eta} \right).$$

Now applying the assumption of the Theorem, we obtain

$$0 \geq -n(n+1)(1 - u(p)),$$

which implies that $u(p) \leq 1$. This yields the desired conclusion.

Thus we deal with:

General case: $u(z) \leq 1$ *for every* $z \in B^n$, *even when* u *does not attain its maximum anywhere on* B^n.

In order to establish such conclusion let q be an arbitrary point of B^n. Let $\|q\| = r_0$.

Let r be an arbitrary constant with $r_0 < r < 1$. Denote by $B^n(0, r)$ the open ball of radius r centered at the origin 0. Let $\varphi_r(z) = z/r$ for every $z \in B^n(0; r)$, and denote by $g_r := \varphi_r^* g$. Recall that

$$g_{ij}\big|_z = \frac{1}{(1 - \|z\|^2)^2} \left((1 - \|z\|^2)\delta_{ij} + \bar{z}_i z_j \right).$$

Using some special features of the Poincaré metric of the unit ball—for instance, it is *Kähler-Einstein* (meaning that the metric is proportional to its Ricci tensor) —Lu introduced the exhaustion function such as

$$v_r(z) = (n + 1) \log \frac{1 - \|z\|^2}{r^2 - \|z\|^2}.$$

Denote by $v = \log u$. Then by the Chern-Lu formulae one obtains

$$\Delta(v - v_r) \geq 4n(n + 1)(e^v - e^{v_r})$$

at every point of $z \in B^n(0, r)$. Since the real exponential function $y = e^x$ is strictly increasing for $x \in \mathbb{R}$, we shall consider the set $E = \{z \in B^n(0, r) : v(z) > v_r(z)\}$. Then $\Delta(v - v_r) > 0$ at every point of E. In particular, $v - v_r$ does not attain any local maximum on E. (Note that E is an open set.)

Now, unless E is empty, one must have a sequence $p_j \in E$ such that $\lim_{j \to \infty}(v(p_j) - v_r(p_j)) = \sup_E(v - v_r)$. Since $p_j \in B^n(0, r)$, the sequence $\{p_j\}$ must have a convergent subsequence converging to $p_0 \in \mathrm{cl}(B^n(0, r))$. If $p_0 \in B^n(0, r)$, say, then $v - v_r > 0$ at p_0. Hence $p_0 \in E$ and consequently $v - v_r$ attains maximum on E, which is a contradiction. If $p_0 \notin B^n(0, r)$, then $\|p_0\| = r$. But then $v_r(p_0) = \infty$. Since $v(p_0)$ is bounded, this implies that $\sup_E(v - v_r) = -\infty$. Thus we can conclude that E is empty.

Therefore, $v \leq v_r$ for any point of $B^n(r)$. In particular,

$$u(q) = e^{v(q)} \leq e^{v_r(q)}.$$

Letting $r \nearrow 1$ we see that $u(q) \leq 1$. This completes the proof. $\square$

Of course it is worth reading the original text ([Chern 1968], [Lu 1968]).

Chapter 6

Tamed Exhaustion and Almost Maximum Principle

The generalization of Schwarz's Lemma by Chern and Lu in the preceding chapter gives considerable information regarding how to handle higher dimensional cases. On the other hand, the "shrinking method" was still present and remains practically the same as in the original Riemann surface result of Ahlfors. Thus an effective method replacing this "shrinking" is in order, when a generalization of Schwarz's Lemma needed to treat holomorphic mappings from a general complex Hermitian manifold into another; the shrinking idea will not be available in the general case.

In this chapter, two preparatory results are going to be discussed: (1) we shall prove a generalized Maximum Principle from the viewpoint of Royden's exhaustion function; (2) we shall give an alternative proof to the Almost Maximum Principle by Omori and Yau. ([Omori 1967], [Yau 1975])

At the risk of repeating ourselves excessively, we remark that the entire contents of this chapter are solely Riemannian geometric.

6.1 Tamed Exhaustion

In [Royden 1980], a special type of exhaustion function was introduced. (An *exhaustion function* on a non-compact manifold M is a function $u: M \to \mathbb{R}$ such that $u^{-1}((-\infty, \alpha])$ is compact in M, for every $\alpha \in \mathbb{R}$.) We start with:

Definition 6.1 (Royden). Let M be a Riemannian manifold. A continuous exhaustion function $u : M \to \mathbb{R}$ is called a *tamed exhaustion function* of M, if it satisfies the following two conditions:

    (i) $u \geq 0$.

    (ii) There exists a constant $C > 0$ such that, at every $p \in M$, there exist an open neighborhood V of p and a $\mathcal{C}^2$ smooth function $v\colon V \to \mathbb{R}$ satisfying: $v(p) = u(p)$, $v(x) \geq u(x)$ for any $x \in V$, $\|\nabla v(p)\| \leq C$ and $\Delta v(p) \leq C$. We call such v a *tamed upper supporting function* for u at p.

The existence of such tamed exhaustion shall be established in the following lemma.

Lemma 6.1. *Every complete Riemannian manifold with its Ricci curvature bounded from below admits a tamed exhaustion function.*

Proof. The proof is a direct consequence of the Hessian Comparison Theorem by Greene and Wu ([Greene and Wu 1979]). Let M be a complete Riemannian manifold with dimension m. Let ρ denote its distance. Assume that its Ricci curvature is bounded from below by some negative constant $-c^2$. Fix $x_0 \in M$. Let $r(x) := \rho(x_0, x)$ for every $x \in M$. Then, for every $x \in M \setminus \{x_0\}$, $|\nabla r(x)| = 1$. Before hitting the cut locus of x_0, the function r is smooth and satisfies the estimate

$$\Delta r \leq \frac{m-1}{r} + c\sqrt{m-1}.$$

Let x be a cut point. Then connect x_0 to x be a distance realizing unit speed geodesic, say γ. Consider a geodesic convex open neighborhood U of x, and choose $y \in U \cap \gamma$. Then let $v(z) := r(y) + \rho(y, z)$ for $z \in U$. Then v is smooth ($\mathcal{C}^\infty$) in U, $v(x) = r(x)$, $v(z) \geq r(z)$ for $z \in U$, $\|\nabla v(x)\| = 1$ and $\Delta v(x) \leq \frac{m-1}{\rho(y,x)} + c\sqrt{m-1}$.

Notice that r is a proper function by the Hopf-Rinow Theorem of Riemannian Geometry by the completeness assumption. (See [Cheeger and Ebin 1975] for instance.) Of course, the estimates above are only good away from x_0, but that can easily be taken care of by a small local modification of r near the point x_0. Hence r gives rise to a desired tamed exhaustion function. $\square$

Tamed exhaustion functions can exist even when the Ricci curvature is not bounded from below; this fact turns out to be useful in many cases.

6.2   Almost Maximum Principle

The main utility of the tamed exhaustion function can be seen from the following Generalized Maximum Principle of H. Omori ([Omori 1967]) and

S.T. Yau ([Yau 1975]).

We say that the *Almost Maximum Principle holds for a Riemannian manifold M* if the following property holds:

> *for every C^2 smooth function $f : M \to \mathbb{R}$ that is bounded from above, there exists a sequence $\{p_k\}$ in M such that*
>
> $$\lim_{k \to \infty} \|\nabla f(p_k)\| = 0, \ \limsup_{k \to \infty} \Delta f(p_k) \leq 0, \ \text{and} \ \lim_{k \to \infty} f(p_k) = \sup_M f.$$

Theorem 6.1 (Omori/Yau). *The Almost Maximum Principle holds for any complete Riemannian manifold M with Ricci curvature bounded from below.*

This follows from the following more general statement:

Proposition 6.1 ([Kim and Lee 2007]). *The Almost Maximum Principle holds for any Riemannian manifold that admits a tamed exhaustion function.*

Proof. The proof is essentially the same as the one developed by Omori and also by Yau, but we give details for the sake of completeness. Let u be a tamed exhaustion function. For each integer $k > 0$, consider $f_k(x) = f(x) - u(x)/k$. Since u is an exhaustion $f_k(x) \to -\infty$ as x runs away indefinitely far from a fixed point. Therefore, there exists $p_k \in M$ at which f_k attains its maximum. Now, let v be a tamed upper supporting function for u at p_k. Then $f(x) - \frac{1}{k}v(x)$ attains its local maximum at p_k. Hence one immediately has

$$\nabla f(p_k) - \frac{1}{k}\nabla v(p_k) = 0$$

and

$$\Delta f(p_k) - \frac{1}{k}\Delta v(p_k) \leq 0.$$

Therefore, $\|\nabla f(p_k)\| \leq C/k$ and $\Delta f(p_k) \leq C/k$.

Finally, it remains to check whether $f(p_k)$ converges to $\sup_M f$ as $k \to \infty$. Let $\epsilon > 0$. Then there exists $p \in M$ such that $f(p) > \sup_M f - \epsilon/2$. Now choose k sufficiently large that $2u(p) \leq k\epsilon$. Then it follows that

$$f(p_k) \geq f(p_k) - \frac{1}{k}u(p_k) \geq f(p) - \frac{1}{k}u(p) \geq \sup_M f - \epsilon.$$

The result follows immediately.   □

It is worth noting that a tamed exhaustion function can be guaranteed to exist even if there is no lower bound for the Ricci tensor, as long as the Ricci curvature does not tend to negative infinity too fast. Thus generalized sufficient conditions for the Almost Maximum Principle are known. Most notable work seems [Ratto, Rigoli and Setti 1995].

On the other hand, some curvature condition is necessary in order for the Almost Maximum Principle to hold in general. We shall give several examples starting with the example presented by Omori himself.

Example 6.1 ([Omori 1967]). The underlying manifold is the Euclidean plane, i.e., $M = \mathbb{R}^2$. The Riemannian metric we use is given in polar coordinate system (r, θ) by

$$ds^2 = dr^2 + g(r, \theta)\, d\theta^2$$

with the C^∞ positive function

$$g(r, \theta) = \begin{cases} r & \text{if } 0 \leq r < \tfrac{1}{2} \\ \exp\left(\displaystyle\int_1^r \frac{(1+t^2)^2}{t}\, dt\right) & \text{if } r > 1. \end{cases}$$

Let

$$f(r, \theta) = \frac{r^2}{1 + r^2}.$$

Then

(i) $f \colon M \to \mathbb{R}$ is C^∞ on M and $f < 1$ everywhere.
(ii) $f(r)$ approaches its supremum as $r \to \infty$.
(iii) $|\nabla f(r, \theta)| \to 0$ as $r \to \infty$.
(iv) $\Delta f(r, \theta) \geq \tfrac{1}{2}$ as $r \to \infty$.
(v) The curvature $K(r, \theta) \sim -\tfrac{1}{4} r^6$ as $r \to \infty$.

This justifies the necessity of the curvature condition in the Almost Maximum Principle (Theorem 6.1 and Proposition 6.1 above).

We include some of the details for the computation. In this example, regard the coordinate functions ordered such as $r = x_1$ and $\theta = x_2$. Since the functions f and g above are independent of the θ-variable, we shall simply write them as $f(r)$, $g(r)$. The covariant derivative ∇ (= Levi-Civita connection) can be computed directly. The Christoffel symbols Γ_{ij}^k are by definition the coefficients in the following formula

$$\nabla_{\frac{\partial}{\partial x_i}} \frac{\partial}{\partial x_j} = \sum_k \Gamma_{ij}^k \frac{\partial}{\partial x_k}.$$

If we employ the usual notation $g_{ij} = \langle \frac{\partial}{\partial x_i}, \frac{\partial}{\partial x_j} \rangle$, then $g_{11} = 1, g_{12} = 0 = g_{21}, g_{22} = g(r)$. The standard formulas of differential geometry give

$$\Gamma_{11}^1 = \Gamma_{11}^2 = \Gamma_{12}^1 = \Gamma_{21}^1 = \Gamma_{22}^2 = 0,$$

$$\Gamma_{12}^2 = \Gamma_{21}^2 = \frac{1}{2} \cdot \frac{g'(r)}{g(r)}, \text{ and } \Gamma_{22}^1 = -\frac{1}{2} g'(r).$$

For the Laplacian, we use the second covariant derivative of the function f which is defined to be

$$\nabla^2 f(X, Y) = X(Yf) - (\nabla_X Y)f.$$

If we denote by

$$L_{ij} = \nabla^2 f \left(\frac{\partial}{\partial x_i}, \frac{\partial}{\partial x_j} \right),$$

then

$$L_{11} = f''(r), \quad L_{12} = 0 = L_{21}, \quad L_{22} = \frac{1}{2} g'(r) f'(r).$$

Then the Laplacian is

$$\Delta f := \text{trace } \nabla^2 f = \sum_{i,j=1}^{2} g^{ij} L_{ij} = f''(r) + \frac{1}{2} \frac{g'(r)}{g(r)} f'(r),$$

where g^{ij} is the (i, j)-th entry of the inverse matrix to (g_{ij}); indeed

$$g^{11} = 1, \quad g^{12} = 0 = g^{21}, \quad g^{22} = \frac{1}{g(r)}.$$

Now the reason for the choice of $g(r)$ for large values for r becomes apparent: it satisfies

$$\frac{g'(r)}{g(r)} f'(r) = 2.$$

Since $f''(r) \to 0$ as $r \to \infty$, we see immediately that (iv) holds. Checking of other details are left to the reader as an exercise.

This example of Omori illuminates the role of the hypothesis of lower curvature bound in the Almost Maximum Principle, as discussed earlier.

On the other hand one may ask what the "sharp" condition is, the weakest curvature hypothesis that suffices. Proposition 6.1 says that the Almost Maximum Principle holds when the manifold admits a complete metric and a tamed exhaustion function. Hence one cannot help thinking that the condition for the existence of a tamed exhaustion function is a key to further generalization of the almost maximum principle. There have been various studies in this direction.

A tamed exhaustion function exists. The most general condition known up to now (See [Ratto, Rigoli and Veron 1994])is:

$$\mathrm{Ric}_M(\nabla r, \nabla r) \gtrsim -r^2(\log(r))^2(\log(\log(r)))^2 \cdots (\log^{(k)}(r))^2, \qquad r \gg 1$$

where $\log^{(k)}$ denotes the composition of k copies of the log-function. This condition is in a sense almost sharp as the following shows:

Example 6.2. On $\mathbb{R}^2$, consider the Riemannian metric in polar coordinate system by

$$ds^2 = dr^2 + g(r,\theta)d\theta^2,$$

where the smooth (C^∞) function g is defined to be

$$g(r,\theta) = \begin{cases} r^2 & \text{if } 0 \leq r < 1 \\ r^{2+2\epsilon}e^{2r^{2+\epsilon}} & \text{if } r > 3, \end{cases}$$

for some positive constant ϵ. Let

$$f(r,\theta) = \int_0^r g(s)^{-1/2}\left(\int_0^s \sqrt{g(t)}dt\right) ds.$$

Then:

(1) f is C^∞ smooth on M and bounded from above.
(2) $\Delta f(r,\theta) = 1$.
(3) The curvature $K(r,\theta) \sim -(2+\epsilon)^2 r^{2+2\epsilon}$ as $r \to \infty$.

Checking the detail is routine possibly except the boundedness of f. The boundedness can be obtained as follows: Since

$$\int_3^s t^{1+\epsilon}\, e^{t^{2+\epsilon}}\, dt \leq s^\epsilon \int_3^s t\, e^{t^2 s^\epsilon}\, dt \leq \frac{1}{2}e^{s^{2+\epsilon}}$$

one has

$$\sup_{r\to\infty} \int_3^r s^{-1-\epsilon}e^{-s^{2+\epsilon}}\left(\int_3^s t^{1+\epsilon}e^{t^{2+\epsilon}}\, dt\right) ds \leq \frac{1}{2}\int_3^\infty s^{-1-\epsilon}ds < \infty,$$

which implies $\sup_M f$ is finite.

Example 6.3. In [Ratto, Rigoli and Setti 1995], an even shaper example is given. The function g in the definition of the Riemannian metric is as follows:

$$g(r,\theta) = \begin{cases} r^2 & \text{if } 0 \leq r < 1 \\ r^2 (\log r)^{2+2\epsilon} e^{2r^2 (\log r)^{1+\epsilon}} & \text{if } r > 3. \end{cases}$$

To check against the almost maximum principle, take the C^∞ function f the same as before. Then one can easily see that f does not satisfy the Almost Maximum Principle. This example is sharper, because the curvature K satisfies

$$K(r,\theta) \sim -c^2 r^2 (\log r)^{2+2\epsilon} \quad \text{as } r \to \infty,$$

featuring a slightly slower rate of curvature decay to $-\infty$.

Chapter 7

General Schwarz's Lemma by Yau and Royden

We are now ready to present the generalizations of Schwarz's Lemma by S.T. Yau and H.L. Royden. The main contribution of Yau's generalization is in that the holomorphic mappings under consideration are from a general complete Kählerian manifold with Ricci tensor bounded from below into a general Hermitian manifold with its bisectional curvature bounded from above. Royden's contribution was that the negative bound (from above) need only be assumed for the holomorphic sectional curvature of the target manifold, a weaker condition than the bisectional curvature bound.

7.1 Generalization by S.T. Yau

One of the most general versions of the differential geometric generalization of Schwarz's Lemma is the following theorem by S.T. Yau, which was also proved (independently) by H.L. Royden ([Royden 1980]).

Theorem 7.1 ([Yau 1978]). *Let (M, g) be a complete Kähler manifold with its Ricci curvature bounded from below by a negative constant $-k$, and let (N, h) be a Hermitian manifold with its holomorphic bisectional curvature bounded from above by a negative constant $-K$. Then every holomorphic mapping $f : M \to N$ satisfies*

$$f^*h \leq \frac{k}{K}\, g.$$

Proof. Start with the Chern-Lu set up $f^*h \leq ug$ and the Chern-Lu formula on u in Theorem 4.3.1 of Section 4.3. Since there is nothing to prove when $u \equiv 0$, we may assume without loss of generality that $\sup_M u > 0$.

The proofs of earlier theorems used the maximum principle for $\log u$ followed by "shrinking methods". But here one does not have any effective shrinking method available. Yau's ingenious discovery relies on an effective(!) functional that replaces the role of the logarithmic function. We shall consider this method carefully.

Consider $\varphi : [0, \infty) \to [0, \infty)$, a C^2 function, with some extra properties that are to be determined later as we continue.

First, require φ to be *monotone decreasing* and *bounded from below*. Then apply the Almost Maximum Principle (Theorem 6.1) of Omori and Yau to $-\varphi \circ u$; namely:

There exists a sequence $\{p_\nu \in M \mid \nu = 1, 2, \ldots\}$ such that

$$\inf_M \varphi \circ u = \lim_{\nu \to \infty} \varphi \circ u(p_\nu), \tag{7.1.1}$$

$$0 = \lim_{\nu \to \infty} \nabla(\varphi \circ u)|_{p_\nu} \tag{7.1.2}$$

and

$$0 \leq \liminf_{\nu \to \infty} \Delta(\varphi \circ u)|_{p_\nu}$$
$$= \liminf_{\nu \to \infty} [\varphi''(u(p_\nu))\|\nabla u|_{p_\nu}\|^2 + \varphi'(u(p_\nu))\Delta u|_{p_\nu}]. \tag{7.1.3}$$

Since φ is going to be chosen to be strictly monotone-decreasing, the condition (7.1.1) implies that

$$\sup_M u = \lim_{\nu \to \infty} u(p_\nu).$$

By the Chern-Lu formula, one has

$$\Delta u = 2\Delta_c u \tag{7.1.4}$$
$$= 2 \sum |a_{\alpha i k}|^2 + \sum a_{\alpha i} \bar{a}_{\alpha j} R_{ij} - \sum a_{\alpha i} \bar{a}_{\beta i} a_{\gamma j} \bar{a}_{\eta j} S_{\alpha \beta \gamma \eta}$$
$$\geq -ku + Ku^2. \tag{7.1.5}$$

Let $\epsilon > 0$ be given. Combining (7.1.3) and (7.1.5) with the above and using $\varphi'(t) < 0$ one sees there exist $N > 0$ such that at every p_ν with $\nu \geq N$

$$2(\varphi'(u))(-ku + Ku^2) + \varphi''(u)\|\nabla u\|^2 > -\epsilon$$

and using (7.1.2)

$$(\varphi'(u))^2 \|\nabla u\|^2 = \|\nabla(\varphi \circ u)\|^2 < \epsilon^2.$$

It follows that

$$(\varphi'(u))^3 \, (-ku + Ku^2) > -\frac{\epsilon}{2}((\varphi'(u))^2 + \epsilon\varphi''(u)).$$

Rewritten, the inequality becomes (where $u > 0$)

$$-k + Ku < \frac{\epsilon}{2} \left(\frac{1}{u|\varphi'(u)|} + \epsilon\frac{\varphi''(u)}{u|\varphi'(u)|^3} \right).$$

Now, we want to choose φ. One may try the function $\varphi(t) = (1+t)^{-a}$ for some $a > 0$. We try to find appropriate value for a so that we may accomplish two goals:

(a) that $\sup_M u$ is bounded.
(b) that $\sup_M u \le k/K$.

If $u(p_\nu)$ diverges to ∞ as $\nu \to \infty$, one immediately notices the following (by a simple calculation): the left-hand side diverges to infinity with the same speed as $u(p_\nu)$, whereas the right-hand side behaves equivalently to $\epsilon(u(p_\nu))^a + \epsilon^2(u(p_\nu))^{2a}$. Thus if we take a so that $0 < a \le 1/2$, then we reach at a contradiction as $\epsilon > 0$ can be chosen arbitrarily small.

Yau's choice for a was $a = 1/2$. Thus we first obtain that $\sup_M u$ is bounded. Moreover, one obtains that

$$u(p_\nu) < \frac{k}{K} + \frac{\epsilon}{2K} \left(\frac{1}{u(p_\nu)|\varphi'(u(p_\nu))|} + \epsilon\frac{\varphi''(u(p_\nu))}{u(p_\nu)|\varphi'(u(p_\nu))|^3} \right).$$

Finally, let $\nu \to \infty$. Then since $\epsilon > 0$ is arbitrary, one arrives at

$$\sup_M u \le \frac{k}{K},$$

as desired. $\qquad\qquad\square$

7.2 Schwarz's Lemma for Volume Element

In the paper [Yau 1978], Yau also presented the following generalized Schwarz's lemma for volume elements:

Theorem 7.2 ([Yau 1978]). *Let M be a complete Kähler manifold with scalar curvature bounded from below by K_1. Let N be another Hermitian manifold with Ricci curvature bounded above by a negative constant K_2.*

Suppose that the Ricci curvature of M is bounded from below and $\dim M = \dim N$. *Then the existence of a non-degenerate holomorphic map f from M into N implies that $K_1 \leq 0$ and*

$$f^* dV_N \leq \frac{K_1}{K_2} dV_M,$$

where dV_M, dV_N are volume elements of M and N, respectively.

This theorem implies an interesting consequence for the Einstein-Kähler metric for bounded pseudoconvex domains constructed in [Cheng and Yau 1980] and in [Mok and Yau 1983]. (See also [Greene-Kim-Krantz 2010], especially Chapter 7.) For a bounded strongly pseudoconvex domain in $\mathbb{C}^n$ with smooth boundary for instance, S.Y. Cheng and S.T. Yau proved that there exists a complete Kähler metric whose Ricci tensor is equal to the negative of the metric itself. Then they showed that this metric, which is called the Cheng-Yau Einstein Kähler metric, for this domain is unique. The uniqueness comes from the above theorem, the volume version of the generalized Schwarz's Lemma.

The argument is simple: If another such metric existed, when one scales it by multiplying a positive constant, the Ricci tensor will be equal to one of the following three: (1) the metric, (2) zero identically, or (3) negative of the metric. Since the identity map is a non-degenerate holomorphic map, Theorem 7.2 tells us that the third case is the only possibility. So we are only to show that the Cheng-Yau metric for the domain with Ricci curvature -1 is unique. Again, Theorem 7.2 implies that their volume forms coincide, inequality running both ways because $K_1 = -1 = K_2$. In coordinates, this means that the determinants of the metric tensors coincide. Thus the complex Hessian of their logarithms must coincide also. But then, the complex Hessian of log of the determinant of the metric tensor is the Ricci tensor (cf., e.g., formula (24) in Page 158, Volume II, [Kobayashi and Nomizu 1969]) in each case. By the Einstein equation which these metrics satisfy, we see now that the metrics coincide! Thus, for each bounded pseudoconvex domain, there can be only one normalized complete Einstein-Kähler metric. The proof that there is one is a deep result using Monge-Ampère equation estimates [Cheng and Yau 1980].

7.3 Generalization by H.L. Royden

We describe Royden's generalization of Schwarz's Lemma. Here, only a negative upper bound for the *holomorphic sectional curvature* is assumed, *a priori* a weaker condition than a negative upper bound for bisectional curvature.

Theorem 7.3 ([Royden 1980]). *Let $f : M \to N$ be a holomorphic mapping from a complete Kähler manifold (M, g) with its Ricci curvature bounded from below by a negative constant $-k$ into a Hermitian manifold (N, h) with its holomorphic sectional curvature bounded from above by a negative constant $-K$. If ν is the maximal rank of the map f, then*

$$f^*h \leq \frac{2\nu}{\nu+1}\frac{k}{K}g.$$

Proof. The proof follows by Yau's generalization of Schwarz's Lemma in the preceding section and a multi-linear algebra technique relating the bound for bisectional curvature and the bound for holomorphic sectional curvature discovered by H.L. Royden which we shall describe now:

Assume that the holomorphic sectional curvature of h bounded from above by the negative constant $-K$. With the notation used above, it suffices show

$$\sum_{i,j,\alpha,\beta,\gamma,\eta} S_{\alpha\beta\gamma\eta}a_{\alpha i}\bar{a}_{\beta i}a_{\gamma j}\bar{a}_{\eta j} \leq -\frac{\nu+1}{2\nu}Ku^2$$

where ν is the rank of df. On the other hand, this inequality follows from the lemma below:

Lemma 7.1 ([Royden 1980]). *Let $\xi_1, \ldots \xi_\nu$ be mutually orthogonal non-zero tangent vectors. Suppose that $S(\xi, \bar{\eta}, \zeta, \bar{\omega})$ is a symmetric "bi-hermitian" form which means that S has the property: $S(\xi, \bar{\eta}, \zeta, \bar{\omega}) = S(\zeta, \bar{\eta}, \xi, \bar{\omega})$. Suppose also that $S(\xi, \bar{\eta}, \zeta, \bar{\omega}) = S(\eta, \bar{\xi}, \omega, \bar{\zeta})$ and $S(\xi, \bar{\xi}, \xi, \bar{\xi}) \leq K\|\xi\|^4$, for all ξ. Then*

$$\sum_{\alpha,\beta} S(\xi_\alpha, \bar{\xi}_\alpha, \xi_\beta, \bar{\xi}_\beta) \leq \frac{1}{2}K\left[\left(\sum_\alpha \|\xi_\alpha\|^2\right)^2 + \sum_\alpha \|\xi_\alpha\|^4\right].$$

If $K \leq 0$, then

$$\sum_{\alpha,\beta} S(\xi_\alpha, \bar{\xi}_\alpha, \xi_\beta, \bar{\xi}_\beta) \leq \frac{\nu+1}{2\nu}K\left(\sum_\alpha \|\xi_\alpha\|^2\right)^2.$$

Proof. Consider $\mathbb{Z}_4^\nu \ni A = (\epsilon_1, \ldots, \epsilon_\nu)$, where $\epsilon_\alpha \in \{1, -1, \sqrt{-1}, -\sqrt{-1}\}$. Let

$$\xi_A = \sum \epsilon_\alpha \xi_\alpha.$$

Then $\|\xi_A\|^2 = \sum \|\xi_\alpha\|^2$, and so

$$S(\xi_A, \bar{\xi}_A, \xi_A, \bar{\xi}_A) \le K\|\xi_A\|^4 \le K\left(\sum_\alpha \|\xi_\alpha\|^2\right)^2.$$

Hence

$$K\left(\sum_\alpha \|\xi_\alpha\|^2\right)^2 \ge \frac{1}{4^\nu} \sum_{A \in \mathbb{Z}_4^\nu} S(\xi_A, \bar{\xi}_A, \xi_A, \bar{\xi}_A)$$

$$= \frac{1}{4^\nu} \sum_{A \in \mathbb{Z}_4^\nu} \epsilon_\alpha \epsilon_\beta \epsilon_\gamma \epsilon_\delta \, S(\xi_\alpha, \bar{\xi}_\beta, \xi_\gamma, \bar{\xi}_\delta)$$

$$= \sum_\alpha S(\xi_\alpha, \bar{\xi}_\alpha, \xi_\alpha, \bar{\xi}_\alpha)$$

$$+ \sum_{\alpha \ne \gamma} S(\xi_\alpha, \bar{\xi}_\alpha, \xi_\gamma, \bar{\xi}_\gamma) + S(\xi_\alpha, \bar{\xi}_\gamma, \xi_\gamma, \bar{\xi}_\alpha).$$

By the symmetry of S, we have

$$\sum_\alpha S(\xi_\alpha, \bar{\xi}_\alpha, \xi_\alpha, \bar{\xi}_\alpha) + 2 \sum_{\alpha \ne \gamma} S(\xi_\alpha, \bar{\xi}_\alpha, \xi_\gamma, \bar{\xi}_\gamma) \le K\left(\sum_\alpha \|\xi_\alpha\|^2\right)^2.$$

Add $\sum_\alpha S(\xi_\alpha, \bar{\xi}_\alpha, \xi_\alpha, \bar{\xi}_\alpha)$ to both sides and use upper bound condition of S to deduce

$$2 \sum_{\alpha, \gamma} S(\xi_\alpha, \bar{\xi}_\alpha, \xi_\gamma, \bar{\xi}_\gamma) \le K\left[\left(\sum_\alpha \|\xi_\alpha\|^2\right)^2 + \sum_\alpha \|\xi_\alpha\|^4\right].$$

Suppose $K \le 0$. Since $(\sum_\alpha \|\xi_\alpha\|^2)^2 \le \nu \sum_\alpha \|\xi_\alpha\|^4$, we obtain

$$\sum_{\alpha, \gamma} S(\xi_\alpha, \bar{\xi}_\alpha, \xi_\gamma, \bar{\xi}_\gamma) \le \frac{\nu + 1}{2\nu} K\left(\sum_\alpha \|\xi_\alpha\|^2\right)^2,$$

as desired. $\square$

Chapter 8

More Recent Developments

In Ahlfors' generalization of Schwarz's Lemma, the completeness of the Poincaré metric of the disc played an important role. The completeness of the metric of the source manifold continued to play an essential role in all the generalizations of Schwarz's Lemma (after Ahlfors') which we introduced up to now. It is natural to ask how Schwarz's Lemma can be reformulated in the case when the source disc is equipped with an *incomplete* metric. Osserman answered this question for the holomorphic maps in complex dimension 1, from a geodesic disc into another (cf. [Osserman 1999a], [Osserman 1999b]). The first purpose of this chapter is to present a brief survey of Osserman's work.

The strict negativity assumption on the curvature of the target manifold is another aspect of Ahlfors' generalization of Schwarz's Lemma and further generalizations. Again, there is a question of whether the condition of a negative upper-bound can be relaxed. This was investigated earlier also (cf. [Greene and Wu 1979], e.g.); we shall briefly survey on these results concerning the case of non-positively curved target Riemann surfaces, by Troyanov and by Ratto-Rigoli-Véron. We shall not, however, go too deeply into the full detail of the expositions, nor to attempt to cover the wide collection of further contributions that are related. We stop at the point at which we seem to have provided a "lead" toward this subject of active research.

8.1 Osserman's Generalization

The mappings to consider in this section are from a real 2-dimensional disc, say $\widehat{D}$, into another 2-dimensional disc D.

If M is a surface with Riemannian metric ds^2 and if $p \in M$, a geodesic disc D of (Riemannian) radius ρ_0 (centered at p) by definition the image, by the exponential map, of the disc of radius ρ_0 centered at the origin in the tangent space T_pM (with respect to the Riemannian metric at p), when ρ_0 is small enough that this exponential map is a diffeomorphisms, i.e., $\rho_0 \leq$ the injectivity radius at p. Then one has the usual representation of the metric on D in terms of geodesic polar coordinates as follows:

$$ds^2 = d\rho^2 + G(\rho, \theta)^2 d\theta^2,$$

where $\rho(q)$ is the distance to $q \in D$ from the center p of the disc D, and where the positive smooth function $G \colon D \to \mathbb{R}$ satisfying

$$G(0, \theta) = 0, \quad \frac{\partial G}{\partial \rho}(0, \theta) = 1, \quad G(\rho, \theta) > 0$$

for $0 < \rho < \rho_0$.

To introduce the key comparison lemma of Osserman, we need notation. Let M and $\widehat{M}$ be surfaces with Riemannian metrics ds^2 and $d\hat{s}^2$, respectively. Let D be a geodesic disc centered at p in M and let $\widehat{D}$ be also a geodesic disc in $\widehat{M}$ centered at $\hat{p}$, respectively. Write the metrics in the respective geodesic polar coordinates:

$$ds^2 = d\rho^2 + G(\rho, \theta)^2 d\theta^2 \quad \text{and} \quad d\hat{s}^2 = d\hat{\rho}^2 + \widehat{G}(\hat{\rho}, \theta)^2 d\theta^2.$$

Denote by K and $\widehat{K}$ the (Gauss) curvatures for ds^2 and $d\hat{s}^2$, respectively. At this juncture, we cite the following lemma, which is actually a corollary to the Greene-Wu Hessian comparision theorem ([Greene and Wu 1979] for the full version):

Lemma 8.1 (Laplacian Comparison). *If $K(y) \leq \widehat{K}(x)$ for all $x \in \widehat{D} \setminus \{\hat{p}\}$ and $y \in D \setminus \{p\}$ satisfying the equality $\hat{\rho}(x) = \rho(y)$, then*

$$\Delta\rho(y) \geq \widehat{\Delta}\hat{\rho}(x)$$

for any such x and y. Here $\widehat{\Delta}$ is the Laplacian with respect to the metric of $\widehat{D}$.

Now we present Osserman's Finite Shrinking Lemma:

Theorem 8.1 ([Osserman 1999b]). *Let $\widehat{M}$ be a Riemann surface equipped with a Hermitian metric $d\,\hat{s}^2$ and let $\widehat{D}$ be a geodesic disc of radius ρ_1. Also suppose that D is a geodesic disc of radius ρ_2 in another Riemann surface, say M, equipped with a Hermitian metric ds^2. Assume that $d\hat{s}^2$ on $\widehat{D}$ is rotationally symmetric, that is, for a geodesic polar coordinate system $(\hat{\rho}, \theta)$ at the center*

$$d\hat{s}^2 = d\hat{\rho}^2 + \widehat{G}(\hat{\rho})^2 d\theta^2, \quad 0 \leq \hat{\rho} < \rho_1. \tag{8.1.1}$$

Let $f\colon \widehat{D} \to D$ be a holomorphic map from $\widehat{D}$ into a geodesic disc D, with center p at the image $f(\hat{p})$ under f of the center $\hat{p}$ of $\widehat{D}$. If $\rho_2 \le \rho_1$, and if

$$K(y) \le \widehat{K}(x) \text{ for any } x, y \text{ with } \rho(y) = \hat{\rho}(x) \tag{8.1.2}$$

then

$$\rho(f(x)) \le \hat{\rho}(x) \text{ for all } x \text{ in } \widehat{D}.$$

Unlike the preceding generalizations of Schwarz's Lemma, this theorem contains seemingly several more restrictions in its hypothesis. This is due to the possible incompleteness of the metric on $\widehat{D}$. Before beginning the proof, we illustrate that such restrictions are indeed essential, especially the requirement $\rho_2 \le \rho_1$, through the following two simple examples.

Example 8.1. Let $\widehat{D}$ be the open unit disc $\{z \in \mathbb{C}\colon |z| < 1\}$ equipped with the standard Euclidean metric (*incomplete*, with curvature 0) and D be the same unit disc but equipped with the Poincaré metric (complete!) with curvature -1. Consider the identity map (clearly holomorphic!) from $\widehat{D}$ to D, then this map is distance increasing, even though the curvature of the image disc is less than the curvature of the source disc. Notice here that the condition $\rho_2 \le \rho_1$ was violated, because $\rho_1 = 1$ and $\rho_2 = +\infty$.

Example 8.2. This time, we change the setting slightly. Let $\widehat{D}$ be the unit disc in $\mathbb{C}$ equipped with Euclidean metric as before. But then we take D as the same unit disc but equipped with the Hermitian metric $ds^2 = \frac{4}{(4-|z|^2)^2}|dz|^2$. This metric is complete, with curvature -4, for the disc in $\mathbb{C}$ with radius 2 (centered at the origin) but not complete when restricted to D. Consider again the identity map $\iota\colon \widehat{D} \to D$. It is easy to check that

$$\rho(\iota(z)) = \int_0^{|z|} \frac{2}{4-t^2} dt = \frac{1}{2} \log \frac{2+|z|}{2-|z|} \le |z| = \hat{\rho}(z)$$

for any $z \in \widehat{D}$. Thus the identity map shrinks the distance from the origin. Notice that the condition on the radii of the geodesic discs is met; the radius of D with respect to ds^2 is $(1/2)\log 3$ which is less than the (Euclidean) radius 1 of $\widehat{D}$.

Proof of Theorem 8.1. Osserman's proof is as follows: Since $d\hat{s}^2$ on $\widehat{D}$ is rotationally symmetric, we may assume without loss of generality that

$$\widehat{D} = \{z \in \mathbb{C}\colon |z| < R\}$$

and

$$d\hat{s}^2 = \hat{\lambda}(r)^2 |dz|^2, \quad |z| < R \le \infty,$$

where $|dz|^2 = dr^2 + r^2 d\theta^2$ denotes the Euclidean metric of $\mathbb{C}$ in (Euclidean) polar coordinate system (r, θ). Comparing it with (8.1.1) we have

$$d\hat{\rho} = \hat{\lambda}(r) dr.$$

Hence $\hat{\rho}$ can be expressed in terms of the Euclidean polar coordinate system (r, θ) as follows:

$$\hat{\rho} = h(r) =: \int_0^r \hat{\lambda}(t) dt,$$

whenever $0 \le r < R$. Note that $r \to h(r)$ is a strictly increasing function (real-valued with a single real variable r), satisfying $h(R) = \rho_1$. Thus it has the inverse function H satisfying $r = H(\hat{\rho})$, whenever $0 \le \hat{\rho} < \rho_1$. Of course $H(\rho_1) = R$.

We briefly summarize Osserman's proof: Consider $\dfrac{H(\rho(f(z)))}{H(\hat{\rho}(z))}$. This function turns out to be subharmonic on $\widehat{D}$. Thus the weak maximum principle (Corollary 1.2) implies that

$$\sup_{z \in \widehat{D}} \frac{H(\rho(f(z)))}{H(\hat{\rho}(z))} \le \sup_{z \in \partial\widehat{D}} \frac{H(\rho(f(z)))}{H(\hat{\rho}(z))}.$$

It also turns out that the right-hand side is less than or equal to 1. The monotone increasing property of H then yields the desired inequality

$$\rho(f(z)) \le \hat{\rho}(z).$$

For detail, see the rest of the arguments.

The argument showing the subharmonicity of the function $\dfrac{H(\rho(f(z)))}{H(\hat{\rho}(z))}$ is as follows: When the metric ds^2 is given in geodesic polar coordinate system (ρ, θ) such as

$$ds^2 = d\rho^2 + G(\rho, \theta)^2 d\theta^2,$$

the Laplacian of a function $\varphi(\rho)$ (independent of the θ-variable) is given by

$$\Delta\varphi = \varphi''(\rho) + \frac{\partial \log G}{\partial \rho} \varphi'(\rho).$$

In particular,

$$\Delta\rho = \frac{\partial \log G}{\partial \rho} = \frac{1}{G}\frac{\partial G}{\partial \rho}.$$

This implies

$$\Delta \varphi = \varphi''(\rho) + \Delta \rho \, \varphi'(\rho). \tag{8.1.3}$$

By the Greene-Wu Hessian comparison theorem (See Lemma 8.1 for our purpose)

$$\Delta \rho|_{\rho=c} \geq \Delta \hat{\rho}|_{\hat{\rho}=c} \quad \text{for } 0 < c < \rho_2.$$

Since $H' > 0$, (8.1.3) and the definition of H imply

$$\Delta \log H(\rho)|_{\rho=c} \geq \Delta \log H(\hat{\rho})|_{\hat{\rho}=c} = \Delta \log |z|.$$

Since $d\hat{s}^2$ is proportional to Euclidean metric (i.e., it is a conformal metric) and since $\log |z|$ is harmonic function in the usual sense (i.e., with respect to Euclidean metric), the right-hand side of the inequality is equal to zero. Since f is holomorphic,

$$\Delta_z \log H(\rho(f(z))) \geq 0,$$

whenever $\rho(f(z)) \neq 0$. (Here Δ_z represents the standard Euclidean Laplacian.) Let

$$u(z) = \log \frac{H(\rho(f(z)))}{|z|}, \quad 0 < |z| < R.$$

Then u is subharmonic on

$$D' = \widehat{D} \setminus \{z \colon z = 0 \text{ or } \rho(f(z)) = 0\}.$$

Recall that $\widehat{D} = \{z \in \mathbb{C} \colon |z| < R\}$. Now we need to understand the behavior of u on $\widehat{D} \setminus D'$ in order to apply maximum principle. Note that $u \to -\infty$ as $\rho(f(z)) \to 0$. To analyze $u(z)$ near $z = 0$ we represent f by $w = F(z)$ in terms of a local thermal coordinate w near $f(0)$, with $w = 0$ at $f(0)$. Now we claim that

$$\lim_{z \to 0} \frac{H(\rho(f(z)))}{|z|} = \frac{\lambda(0)}{\hat{\lambda}(0)} |F'(0)|,$$

where $ds^2 = \lambda^2(w)|dw|^2$. To verify the claim, observe that

$$\rho(w) = \int_0^1 \| \frac{d}{dt}(t \to tw) \|_{ds^2} \, dt$$

$$= \int_0^1 \lambda(tw)|w| dt$$

$$= \lambda(0)|w| + O(|w|^2).$$

Thus, near $z = 0$, we see that

$$\rho(f(z)) = \rho(F(z)) = \lambda(0)|F'(0)||z| + O(|z|^2).$$

Notice that $H(\hat{\rho}) = 1/\hat{\lambda}(0)\,\hat{\rho} + O(\hat{\rho}^2)$ near $z = 0$. The claim follows.

The claim yields that, if $F'(0) = 0$, then we have $u(z) \to -\infty$ as $z \to 0$. The weak maximum principle says that u attains its maximum on the boundary of $\hat{D}$. If $F'(0) \neq 0$, apply the same argument to $u_\epsilon = u(z) + \epsilon \log |z|$ for any $\epsilon > 0$, and then let $\epsilon \to 0$ to obtain the same result for u. In either case, we have

$$\log \frac{H(\rho(f(z)))}{|z|} = u(z) \leq \limsup_{|z| \to R} u(z) \leq \log \frac{H(\rho_2)}{R}.$$

Since $H(\rho_1) = R$ and since H is increasing, we see that

$$H(\rho(f(z))) \leq \frac{H(\rho_2)}{H(\rho_1)}|z| \leq |z|.$$

Applying $h \; (= H^{-1})$, we have

$$\rho(f(z)) \leq h(|z|) = \hat{\rho}(z),$$

as desired. This completes the proof.   □

The reader might feel that the conclusion of the theorem seems to assert the distance-decreasing property from the center of each geodesic disc only, and hence the theorem does not seem very general. But this theorem is more general than it appears: as a demonstration we shall see that the above theorem implies Ahlfors' generalization of Schwarz's Lemma as Osserman writes in [Osserman 1999b].

Corollary 8.1 (Ahlfors-Schwarz Lemma). *Let f be a holomorphic map of the unit disc D into a Riemann surface S endowed with a Riemannian metric ds^2 with curvature $K \leq -1$. Then*

$$\mathrm{dist}_S(f(z_1), f(z_2)) \leq \mathrm{dist}_D(z_1, z_2),$$

where dist_S, dist_D are the distances on S and D induced from the respective Riemannian metrics.

Proof. We may assume with no loss of generality that S is simply connected. If it is not, one simply needs to lift the map f to a holomorphic map $\tilde{f}$ into the universal covering space $\tilde{S}$ of S (which is again a Riemann surface).

Let z_1, z_2 be two arbitrarily chosen points in the unit disc D. Since an isometry of D with respect to the Poincaré metric $d\hat{s}^2$ is a holomorphic automorphism of D (up to a conjugation), we can always replace f by the composition of f with a Poincaré isometry taking 0 to z_1. Consequently we may assume without loss of generality that $z_1 = 0$. Thus it suffices to show that

$$\text{dist}_S(f(0), f(z_2)) \leq \text{dist}_D(0, z_2), \tag{8.1.4}$$

for every $z \in D$.

Set $\hat{\rho}(z) = \text{dist}_D(0, z)$ and $\rho(p) = \text{dist}_S(f(0), p)$. The inclusion relation restriction required in the hypothesis of Theorem 8.1—that a geodesic disc centered at $f(0)$ whose radius is not greater than 1 includes $f(D)$—is not automatic in general. So choose r_0 such that

$$|z| < r_0 < 1.$$

Let $\rho_0 = \max_{|z| \leq r_0} \rho(f(z))$. Since S is simply connected and $K < 0$, there exists a global geodesic coordinate system on the disc $D_{\rho_0} = \{p \in S : \rho(p) < \rho_0\}$; this follows by the Cartan-Hadamard theorem in Riemannian geometry (cf., e.g., [Cheeger and Ebin 1975]). Let $\tilde{f}(\zeta) = f(r_0\zeta) : \{\zeta \in \mathbb{C} : |\zeta| < 1\} \to D_{\rho_0}$. Let $d\tilde{s}^2$ be the Poincaré metric on $\{\zeta \in \mathbb{C} : |\zeta| < 1\}$ and $\tilde{\rho}$ the distance to the origin with respect to $d\tilde{s}^2$. Since $d\tilde{s}^2$ is complete on $\{\zeta \in \mathbb{C} : |\zeta| < 1\}$, there exists r_1 such that $|z_2| < r_1 < r_0$ and

$$\rho_1 = \tilde{\rho}\left(\frac{r_1}{r_0}\right) \geq \rho_0.$$

Applying Theorem 8.1 (the finite-shrinking-lemma) to the holomorphic mapping

$$\tilde{f} : \{\tilde{\rho} < \rho_1\} = \{|\zeta| < r_1/r_0\} \to D_{\rho_0}$$

we have $\rho(\tilde{f}(\zeta)) \leq \tilde{\rho}(\zeta)$ for $|\zeta| < r_1/r_0$. In particular, the inequality holds for $\zeta = z_2/r_0$. Thus

$$\rho(f(z_2)) = \rho(\tilde{f}(z_2/r_0)) \leq \tilde{\rho}(z_2/r_0).$$

Now let r_0 approach 1. It follows that $d\tilde{s} \to d\hat{s}$ and $\tilde{\rho}(z_2/r_0) \to \hat{\rho}(z_2)$, as desired. $\qquad\square$

8.2 Schwarz's Lemma for Riemann Surfaces with $K \leq 0$

There are several more and significant generalizations of Schwarz's Lemma. But since there are too many to handle in a set of short lecture notes, we decided to be content with introducing only a few more of them in this section.

One may consider the case when the target manifold is equipped with a Riemannian metric whose curvature is only non-positive. This is again one of the (many) cases that are not covered at all (or, at least not explicitly) by the theorems introduced in this lecture note up to this point. Recall, for instance, that Yau's generalization of Schwarz's Lemma demands that the target manifolds have their bisectional curvatures bounded from above by negative constants. A simple-minded action such as replacing the upper bound for the curvature of the target manifold by 0 will not do; the proof-arguments using the almost maximum principle and the Chern-Lu formula are no longer valid.

Thus the following variation of Schwarz's Lemma by Troyanov which deals with the case in which the target Riemann surface has non-positive curvature is new and worth mentioning.

Theorem 8.2 ([Troyanov 1991]). *Let S_1 be a smooth, complete, connected Riemannian surface equipped with a metric g_1 whose curvature K_1 is bounded from below by some constant. Let S_2 be any smooth Riemannian surface with a metric g_2 and its curvature K_2. If $f : (S_1, g_1) \to (S_2, g_2)$ is a conformal mapping such that*

(1) $K_2 \circ f \leq 0$,
(2) $K_2 \circ f(p) \leq K_1(p)$ *for all* $p \in S_1$,
(3) $K_2 \circ f < -a < 0$ *on the complement of some compact subset of* S_1, *and*
(4) $K_2 \circ f$ *is not identically zero,*

then $f^* g_2 \leq g_1$.

Proof of Theorem 8.2 in a Special Case. A mapping f is said to be *singular* at $p \in S_1$, if there is a complex local coordinate z centered at p such that, for a continuous function v and a real number $\beta > -1$,

$$f^* g_2 = e^{2v(z)} |z|^{2\beta} |dz|^2.$$

We shall only discuss the proof of the (simpler) case when f is non-singular at every point of S_1. Troyanov's original proof of course includes the case when f allows singular points. While we refer the reader to [Troyanov 1991] for complete detail, the discussion in the nonsingular case gives the reader some flavor of this new argument.

The non-singularity of f amounts to

$$f^* g_2 = e^{2u} g_1.$$

The goal here is to establish the estimate $u \leq 0$. Set $P = \{p \in S_1 : u(p) \geq 0\}$. Applying the Chern-Lu formula to $f^*g_2 = e^{2u}g_1$, we see that

$$\Delta u = 2\Delta_c \log(e^{2u}) = K_1 - e^{2u}K_2.$$

Using the comparison condition on $K_2 \circ f$ and K_1, we have

$$-\Delta u = (K_2 \circ f)(e^{2u} - 1) + (K_2 \circ f) - K_1$$
$$\leq 0 \quad \text{on } P.$$

Thus u is subharmonic on P. We now divide the remaining arguments into subcases:

<u>Case 1</u>. *u attains its maximum and $\partial P \neq \emptyset$.* By the maximum principle u has its maximum on ∂P and $u = 0$ identically.

<u>Case 2</u>. *u attains its maximum and $\partial P = \emptyset$.* We must have either $P = \emptyset$ or $P = S_1$. In the latter case u is constant. Computing K_1 and K_2, we deduce that $u = 0$ identically.

<u>Case 3</u>. *u does not attain its maximum.* Suppose $P \neq \emptyset$. Then there exists $\eta > 0$ such that $u(x) > \eta$ for some $x \in S_1$. Apply the almost maximum principle to $\varphi \circ u$, where $\varphi(t) = (1 + e^{-t})^{-1}$. Note that $\sup_{S_1} \varphi \circ u < \infty$. Take $\delta > 0$ so that it satisfies the inequalities $\delta < \sup_{S_1} \varphi \circ u - (1 + e^{-\eta})^{-1}$ and $\epsilon > 0$ such that

$$4\epsilon \leq \frac{a \sinh(\eta)}{1 + 2\sinh(\eta)}.$$

Choose a compact set $N \subset S_1$ such that $K_2 \circ f < -a$ on $S_1 \setminus N$. Then there exists a point $x_{\delta,\epsilon} \in S_1 \setminus N$ at which

$$\Delta(\varphi \circ u) < \epsilon, \quad |\nabla(\varphi \circ u)|^2 < \epsilon$$

and

$$(\varphi \circ u) > \sup_{S_1}(\varphi \circ u) - \delta.$$

Note that $u(x_{\delta,\epsilon}) > \eta$. A direct computation shows that

$$\varphi'(u)\Delta u = \Delta(\varphi \circ u) - \frac{\varphi''(u)}{(\varphi'(u))^2}|\nabla(\varphi \circ u)|^2.$$

We also have

$$\frac{e^{-u}}{(1 + e^{-u})^2}\Delta u < \epsilon(1 + 2\sinh(u)) \quad \text{at } x_{\delta,\epsilon},$$

and $(1 + e^{-u})^2 \leq 4$ at $x_{\delta,\epsilon}$. Since $\Delta u = K_1 - (K_2 \circ f)e^{2u}$, we have

$$(K_2 \circ f)e^u - K_1 e^{-u} \geq -4\epsilon(1 + 2\sinh(u)).$$

Divide by $(K_2 \circ f)$. Since $K_2 \circ f \leq K_1$ and $K_2 \circ f < -a$, we see that

$$e^u - e^{-u} \leq e^u - \frac{K_1}{K_2 \circ f} e^{-u} \leq -\frac{4\epsilon(1 + 2\sinh(u))}{K_2 \circ f} \leq \frac{4\epsilon(1 + 2\sinh(u))}{a}.$$

This yields that

$$4\epsilon \geq \frac{2a\sinh(u)}{1 + 2\sinh(u)} > \frac{2a\sinh(\eta)}{1 + 2\sinh(\eta)}.$$

(Here we used that $\sinh(t)/(1 + 2\sinh(t))$ is an increasing function and $u(x_{\delta,\epsilon}) > \eta$.) This, however, contradicts the choice of ϵ. Thus $P = \emptyset$ and $u < 0$ on S_1 or $P = \{u = 0\}$ and $u \leq 0$ on S_1. This establishes the case as desired. $\qquad\square$

Remark 8.1. Theorem 8.2 implies that, under the same curvature conditions, all holomorphic mappings between oriented Riemann surfaces satisfy the distance decreasing property.

Troyanov's version of Schwarz's Lemma has an application to study of the following problem (sometimes called the Berger-Nirenberg problem) of prescribing the curvature on a Riemann surface.

Problem 8.1. Let (S, g) be a Riemann surface of finite topological type. For a given function $K : S \to \mathbb{R}$, find a metric h on S with curvature K, which is conformal to and conformally quasi-isometric to g, i.e., $h = e^{2u}g$ for some bounded function u.

For the compact case, [Kazdan and Warner 1974] is worth reading. In the non-compact case, for non-positively curved cases, results are known from work of Sattinger and Ni (cf. [Sattinger 1972], [Ni 1982]). If the Riemann surface (S, g) is complete with curvature bounded from above by a negative constant then S is conformally equivalent to the Poincaré disc by the Ahlfors-Schwarz Lemma. Considering the problem of prescribing the curvature for the Poincaré disc, several sufficient conditions on asymptotic behavior of the curvature near the boundary were founded (cf., e.g., [Aviles and McOwen 1985], [Bland and Kalka 1986]). We include only the statements:

Theorem 8.3 ([Bland and Kalka 1986]). Let (D, g) be the Poincaré disc. If a smooth function $K : D \to \mathbb{R}$ satisfies that $K \to 0$ or $K \to -\infty$ near ∂D, then there do not exist any complete metric on D with curvature K, which are conformal and conformally quasi-isometric to g.

This theorem implies that some proper condition for asymptotic behavior of the curvature is necessary in order to solve the problem of prescribing the curvature for Poincaré disc. The following theorem on the other hand presents an affirmative result in a different case:

Theorem 8.4 ([Aviles and McOwen 1985]). *Let (D, g) be the unit disc equipped with the Poincaré metric g. Let $K : D \to \mathbb{R}$ be a smooth function satisfying that $K \leq 0$ on D and $-a^2 \leq K \leq -b^2 < 0$ outside a compact subset of D. Then there exists a unique complete metric on D with curvature K, which is conformal and conformally quasi-isometric to g.*

For the case of complex plane, a result about $\mathbb{C}$ itself can be obtained as a corollary to Theorem 8.2.

Corollary 8.2. *There is no conformal metric g on $\mathbb{C}$ or $\mathbb{C} \backslash \{0\}$ (regardless of its completeness) such that $K_g \leq 0$ and $K_g \leq -a^2 < 0$ outside a compact set.*

The proof is easy: suppose that such a metric g exists on $S = \mathbb{C}$ or $\mathbb{C} \backslash \{0\}$. Let g_c be the constant multiple of the Euclidean metric by c. Then the identity map $id : (S, g_c) \to (S, g)$ is distance decreasing. Since c can be chosen arbitrary, the identity map then must be a constant map. This is impossible and hence the assertion of corollary follows immediately.

We now present another application of Theorem 8.2; it is a generalization of the result of Aviles and McOwen introduced above.

Theorem 8.5 ([Hulin and Troyanov 1992]). *Let S be a connected, open Riemann surface with finite topology which is not biholomorphic to $\mathbb{C}$ or $\mathbb{C} \backslash \{0\}$. Let $K : S \to \mathbb{R}$ be a smooth function satisfying $K \leq 0$ on S, and $-a^2 \leq K \leq -b^2 < 0$ outside a compact subset of S. Then there exists a unique complete conformal metric g on S with curvature K.*

Theorem 8.2 is used to prove the uniqueness of the metric as follows: Suppose that g and h are two such metrics. Apply Theorem 8.2 to the identity map $(S, h) \to (S, g)$ and its inverse. Then $g = h$.

One can see at this juncture that there are two possibilities for further generalization of Troyanov theorem: First, the curvature condition for target surfaces may be relaxed such that theorem includes wider class of nonpositively curved surfaces as target surfaces. Second, one cannot help noticing that the nature of the theorem is essentially theorem of Riemannian geometry. Thus it comes into mind that it should be generalized to higher dimensional Riemannian manifolds. Indeed, for compact Riemannian manifolds, there are earlier investigations by Lichnerowicz, Obata, Yano and Nagano, and others. For complete Riemannian manifold, Yau obtained following theorem using almost maximum principle:

Theorem 8.6 ([Yau 1973]). *Let (M_1, g_1) be a complete connected Riemannian manifold with its sectional curvature bounded from below and its scalar curvature bounded from below by $-k^2$. Let (M_2, g_2) be a connected Riemannian manifold with its scalar curvature bounded from above by $-a^2 < 0$. If $f : (M_1, g_1) \rightarrow (M_2, g_2)$ is a conformal mapping, then*

$$f^* g_2 \leq \frac{k^2}{a^2} g_1.$$

As a generalization of Troyanov theorem along the spirit of Yau's theorem, we introduce theorem by Ratto, Rigoli and Veron [Ratto, Rigoli and Veron 1994].

Let (M_1, g_1) be a connected, complete Riemannian manifold and (M_2, g_2) be a connected Riemannian manifold. Denote by $K_1(x)$ the scalar curvature of (M_1, g_1). Given a diffeomorphism $f : M_1 \rightarrow M_2$, we denote by $K_2(x)$ the scalar curvature of pull-back metric $f^* g_2$. Let $p \in M_1$ and set $r(x) = r_p(x) := \text{dist}_g(x, p)$ for $x \in M_1$. Then we present:

Theorem 8.7 ([Ratto, Rigoli and Veron 1994]). *Suppose that* $\text{Ric}_g \gtrsim (1 + r(x))^{2(1-\gamma)}$ *with* $\gamma \leq 2$. *If* $f : (M_1, g_1) \rightarrow (M_2, g_2)$ *is a conformal mapping such that*

$$K_2(x) \leq \min\{0, K_1(x)\} \qquad \text{for all } x \in M_1$$
$$K_2(x) \lesssim -(1 + r(x))^{-\gamma} \qquad \text{if } r(x) \gg 1,$$

then $f^* g_2 \leq g_1$.

Notice that this theorem implies a generalization of Troyanov's theorem for a class of non-positively curved target surface that is broader than Troyanov's case.

8.3 Final Remarks

Though we do not include in these lecture notes any of the research results on a holomorphic mapping $f \colon M \to N$ between Hermitian manifolds M and N, it seems worth leaving some remarks at his ending stage. As remarked several times earlier, the key ingredients toward establishing and proving a generalization of Schwarz's Lemma seem to be:

(i) To find a suitable function u on M satisfying $f^* h_N \leq u\, h_M$.
(ii) Apply (almost) maximum principle to $\varphi \circ u$ for some appropriate function φ to derive relations such as $\nabla \varphi \circ u \sim 0$ on the gradient, and $\Delta \varphi \circ u \leq \epsilon$ $(\epsilon \to 0)$ on the Laplacian.
(iii) Derive an effective upper bound of u from the relations obtained in the previous step.

As we pointed out several times in these notes, the (almost) maximum principle holds in Riemannian geometry, and hence the Riemannian metric and connection (Levi-Civita) is used in Step (ii). On the other hand the only known method for Step (iii) appears to be the Chern-Lu formulae, which depends on the Hermitian connection. The discrepancies between these two connections and their Laplacians necessarily require additional (sophisticated) conditions. This is what one finds in almost all papers pertaining to generalizations of Schwarz's Lemma for holomorphic mappings between Hermitian manifolds.

However, we remark that such generalizations to Hermitian cases are not just for the sake of theoretical purposes only as one can see from the following result:

Theorem 8.8 ([Seshadri and Zheng 2008]). *If M is the product of two complex manifolds of positive dimensions, then it cannot admit any complete Hermitian metric with bounded torsion and bisectional curvature bounded between two negative constants.*

Bibliography

[Ahlfors 1938] Ahlfors, Lars V., An extension of Schwarz's lemma, *Trans. Amer. Math. Soc.* **43** (1938), no. 3, 359-364.

[Ahlfors 1966] Ahlfors, Lars V., *Complex analysis*: An introduction of the theory of analytic functions of one complex variable, (2nd ed.), McGraw-Hill, New York 1966.

[Aviles and McOwen 1985] Aviles, P. and McOwen, R., Conformal deformations of complete manifolds with negative curvature, *J. Diff. Geometry* **21** (1985) 269-281.

[Bland and Kalka 1986] Bland, J. and Kalka, M., Complete metrics conformal to the hyperbolic disc, *Proc. Amer. Math. Soc.* **97** (1986) 128-132

[Cheeger and Ebin 1975] Cheeger, J. and Ebin, D., Comparison theorems in Riemannian Geometry, *North-Holland* 1975.

[Cheng and Yau 1980] Cheng, Shiu-Yuen and Yau, Shing-Tung, On the existence of a complete Kahler metric on noncompact complex manifolds and the regularity of Fefferman's equation, *Comm. Pure Appl. Math.* **33** (1980), no. 4, 507–544.

[Chern 1968] Chern, S. S., On holomorphic mappings of hermitian manifolds of the same dimension, in "Entire functions and related parts of Analysis (La Jolla, Calif., 1966)", *Proc. Symp. Pure Math.*, Amer. Math. Soc., (1968), 157-170.

[Chern 1979] Chern, S. S., Complex manifolds without potential theory, *Springer-Verlag*, 1979.

[Chern 1989] Chern, S. S., Vector bundles with a connection, *Global differential geometry*, 1–26, MAA Stud. Math., 27, Math. Assoc. America, 1989.

[Gilbarg and Trudinger 1977] Gilbarg, D. and Trudinger, N., *Elliptic partial differential equations of second order* (2nd ed.), Springer-Verlag, Berlin, 1977.

[Grauert and Reckziegel 1965] Grauert, H. and Reckziegel, H., Hermitesche Metriken und normale Familien holomorpher Abbildungen, (German), *Math. Z.* 89 (1965), 108–125.

[Greene 1987] Greene, Robert E., Complex differential geometry, *Differential geometry* (Lyngby, 1985), 228–288, *Lecture Notes in Math.*, **1263**, Springer,

Berlin, 1987.

[Greene, Kim and Krantz 2010] Greene, R. E., Kim, K.-T. and Krantz, S. G., The geometry of complex domains, Birkhäuser-Verlag, 2010.

[Greene and Wu 1979] Greene, R. E. and Wu, H., Function theory on manifolds which possess a pole, *Lect. notes in Math.* 699, Springer-Verlag, 1979.

[Hulin and Troyanov 1992] Hulin, D. and Troyanov, M., Prescribing curvature on open surfaces, *Math. Ann.* **293** (1992), 277-315

[Kazdan and Warner 1974] Kazdan, J. and Warner, F., Curvature functions for compact 2-manifolds, *Ann. Math.* (2) 99 (1974), 14-47.

[Kim and Lee 2007] Kim, K.-T. and Lee, H., On the Omori-Yau almost maximum principle, *J. Math. Anal. Appl.* **335** (2007), 332-340.

[Kobayashi 1967] Kobayashi, S., Distance, holomorphic mappings and Schwarz lemma, *J. Math. Soc. Japan* **19** (1967), 481-485.

[Kobayashi 1967a] Kobayashi, S., Intrinsic metrics on complex manifolds. *Bull. Amer. Math. Soc.* **73** (1967), 347-349.

[Kobayashi 1970] Kobayashi, S., Hyperbolic manifolds and holomorphic mappings, *Marcel-Dekker*, 1970.

[Kobayashi 1998] Kobayashi, S., Hyperbolic complex spaces, *Springer-Verlag*, 1998.

[Kobayashi and Nomizu 1969] Kobayashi, S. and Nomizu, K., Foundations of Differential Geometry, Volume I & II, *Interscience*, 1969.

[Lu 1968] Lu, Y.C., Holomorphic mappings of complex manifolds, *J. Diff. Geom.* **2** (1968), 299-312.

[MathSciNet] MathSciNet, American Mathmematical Society, (http://www.ams.org/mathscinet).

[Mok and Yau 1983] N. Mok and S.-T. Yau, Completeness of the Kähler-Einstein metric on bounded domains and the characterization of domains of holomorphy by curvature conditions, *The mathematical heritage of Henri Poincaré*, Part 1 (Bloomington, Ind., 1980), 41-59, Proc. Sympos. Pure Math., **39**, Amer. Math. Soc., Providence, RI, 1983.

[Ni 1982] Ni, W. M., On the elliptic equation $\Delta u + K(x)e^{2u} = 0$ and conformal metrics with prescribed Gaussian curvatures, *Invent. Math.* **66** (1982) 343-352

[Omori 1967] Omori, H., Isometric immersions of Riemannian manifolds, *J. Math. Soc. Japan.* **19** (1967), 205-214.

[Osserman 1999a] Osserman, R., From Schwarz to Pick to Ahlfors and beyond, *Notices Amer. Math. Soc.* **46** (1999), no. 8, 868-873.

[Osserman 1999b] Osserman, R., A new variant of the Schwarz-Pick-Ahlfors Lemma, *Manuscripta Math.***100** (1999), 123-129.

[Pick 1916] Pick, G., Uber eine Eigenschaft der konformen Abbildung kreisformiger Bereiche, *Math. Ann.* (2) **77** (1916), 1-6.

[Ratto, Rigoli and Veron 1994] Ratto, A., Rigoli, M. and Veron, L., Conformal immersions of complete Riemannian manifolds and extensions of the Schwarz lemma, *Duke Math. J.* **74** (1994), 223-236.

[Ratto, Rigoli and Setti 1995] Ratto, A., Rigoli, M. and Setti, A.G., On the Omori-Yau maximum principle and its application to differential equations and geometry, *J. Func. Anal.* **134** (1995), 486-510.

[Royden 1971] Royden, H. L., *Remarks on the Kobayashi metric*, Several complex variables, II (Proc. Internat. Conf., Univ. Maryland, College Park, Md., 1970), pp. 125–137. Lecture Notes in Math., Vol. 185, Springer, Berlin, 1971.

[Royden 1980] Royden, H.L., The Ahlfors-Schwarz lemma in several complex variables, *Comment. Math. Helv.* **55** (1980), no. 4, 547-558.

[Sattinger 1972] Sattinger, D. H., Conformal metrics in $\mathbb{R}^2$ with prescribed Gaussian curvature, *Indiana Univ. Math. J.* 22 (1972), 1-4.

[Seshadri and Zheng 2008] Seshadri, H. and Zheng, F., Complex product manifolds cannot be negatively curved, *Asian J. Math.* 12 (2008), no. 1, 145–149.

[Troyanov 1991] Troyanov, M., The Schwarz lemma for nonpositively curved Riemann surfaces, *Manuscripta Math.* **72** (1991), 251-256.

[Yau 1973] Yau, S. T., Remarks on conformal transformations, *J. Diff. Geom.* **8** (1973), 369-381.

[Yau 1975] Yau, S.T., Harmonic functions on complete Riemannian manifolds, *Comm. Pure Appl. Math.* **28** (1975), 201-228.

[Yau 1978] Yau, S.T., A general Schwarz lemma for Kähler manifolds, *Amer. J. Math.* **100** (1978), no. 1, 197-203.

Index